T0074657

SAS® Software Companion for Sampling

Design and Analysis

Third Edition

SAS® Software Companion for Sampling

Design and Analysis

Third Edition

Sharon L. Lohr

CRC Press is an imprint of the
Taylor & Francis Group, an **informa** business
A CHAPMAN & HALL BOOK

First edition published 2022
by CRC Press
6000 Broken Sound Parkway NW, Suite 300, Boca Raton, FL 33487-2742

and by CRC Press
2 Park Square, Milton Park, Abingdon, Oxon, OX14 4RN

CRC Press is an imprint of Taylor & Francis Group, LLC

Library of Congress Cataloging-in-Publication Data

ISBN: 978-0-367-74937-8 (hbk)
ISBN: 978-0-367-74851-7 (pbk)
ISBN: 978-1-003-16036-6 (ebk)

DOI: 10.1201/9781003160366

Typeset in SFRM1000 font
by KnowledgeWorks Global Ltd.

Access the Support Material: https://www.routledge.com/9780367748517

To Doug

Contents

Preface

SAS® Software Companion for Sampling: Design and Analysis shows how to use the survey selection and analysis procedures in SAS® software with the examples in the textbook *Sampling: Design and Analysis, Third Edition* (SDA) by Sharon L. Lohr. It is intended to be read in conjunction with SDA and is not a standalone text. The parallel book by Lu and Lohr (2022) shows how to perform the computations for the examples using the R statistical software environment, and could be read together with this book and SDA to learn how to perform the analyses in each software package.

All code and data sets in this book can be downloaded from the website for SDA, which can be reached from either of the following addresses:

`https://www.sharonlohr.com`
`https://www.routledge.com/9780367748517.`

The website also contains additional code, not discussed in this book, that you can adapt for some of the SDA exercises. Many data set formats can be used with SAS software; the SDA website contains the data files in comma-delimited (`.csv`) format as well as in SAS format. The data sets are described in Appendix A.

In this book, I present some of the most frequently used commands for selecting samples and analyzing survey data while working through the examples in Chapters 1–13 of SDA. The software, however, can do much more than analyzing the examples presented in this book. You can find information on advanced capabilities in the *SAS/STAT® User's Guide* (SAS Institute Inc., 2021) and online documentation at `https://support.sas.com`.

The survey design and analysis procedures in SAS/STAT® software provide a powerful platform for selecting samples and performing almost any analysis you would care to do with survey data. The software documentation (SAS Institute Inc., 2021) is easy to read and gives full details on how all calculations are done, along with a multitude of examples that show how the procedures work with survey data sets. The software is available free of charge for students and independent learners; Chapter 1 tells you how to obtain it.

Lewis (2016) presents additional material on the survey analysis procedures in SAS software and is an excellent reference for using SAS software to analyze survey data. Lewis organizes the material differently than this book, devoting a chapter to each procedure that shows it in its full complexity. By contrast, each chapter of this book tells how to use SAS software for the examples in the corresponding chapter of SDA. Chapter 2, for example, shows how to select and compute estimates from simple random samples, Chapter 3 treats stratified random sampling, and so on.

For easy reference, the index at the back of the book gives page numbers for the examples in SDA. To locate the code and output for Example 2.5, for example, look up the subentry "Example 02.05" under "Examples in SDA" in the index. The book also gives code and suggestions for some of the exercises in SDA, and these are listed in the index under "Exercises in SDA."

Each chapter ends with a section on tips and warnings for the procedures and examples discussed in that chapter. These provide ways of avoiding common survey data analysis errors and checking whether you did the analysis correctly.

Finally, a note on the anthropomorphism in this book. I sometimes write that "the program asks for" or "the procedure thinks." Let me clarify that the SAS software procedures discussed herein are computer programs. They are not yet sentient. They do not "think"; they do not "feel"; they do not "want." When I write "the procedure tells you," for example, I mean that the statisticians at SAS Institute, Inc. who developed the program wrote code that produces prespecified text in the log in response to keywords in your program.

Acknowledgments. Many thanks to editor John Kimmel and to the CRC Press production team for their support and help. John encouraged me to write the third edition of SDA, and it was his idea to have separate supplemental books for SAS and R software. It has been an honor and a privilege to be able to work with him.

Tony An and Pushpal Mukhopadhyay provided many helpful comments and suggestions for this book, and answered my multitude of questions with cheerful patience—usually within the same day and often within the same hour. I learned a great deal from our interactions and am profoundly grateful for their help.

Yan Lu performed analyses of the examples in R, providing an independent check of the results in this book. She pointed out where sections in the initial draft could be improved, and her advice greatly improved the book.

And finally, thank you to all of the students in my sampling classes over the years, who tested early versions of the code in this book, provided helpful feedback on how to present the material, and inspired me on this journey.

1

Getting Started

SAS software provides a powerful environment for selecting probability samples and analyzing survey data. The survey analysis procedures will perform almost any analysis you would want to do with survey data. Most analyses can be done using only a few lines of code. The software is designed to allow you to manipulate and manage large data sets (and many survey data sets are quite large, containing tens of thousands or even millions of records), and compute estimates for those data sets using numerically stable and efficient algorithms. And students and independent learners can use a cloud-based version of the software free of charge.

This chapter tells you how to obtain access to the software and introduces you to some basic features. It also shows you how to read data sets into SAS software and save output and graphics that you produce while using it.

Obtaining SAS software. Some institutions have a full license for SAS software, which can be used for all analyses in the book. Another option is to use SAS® OnDemand for Academics, a cloud-based FREE version intended for students and independent learners. Visit `https://www.sas.com/en_us/software/on-demand-for-academics.html` for more information. SAS® OnDemand for Academics contains all of the programs needed to select samples, compute estimates, and graph data for surveys. Instructors can create a course site online and upload data that can be accessed by all students in the course. The "Learn & Support" page provides links to short step-by-step fact sheets on how to get started, video tutorials, frequently asked questions, and more.

Procedures for sample selection and data analysis. Survey data analysis with SAS software is most easily done by writing code to use the procedures (PROCs) that select samples and analyze data from samples. This book will illustrate the use of five of the procedures for survey design and analysis for the examples in *Sampling: Design and Analysis, Third Edition*:

The SURVEYSELECT procedure selects a probability sample from a population—with or without replacement, and with equal or unequal probabilities.

The SURVEYMEANS procedure calculates means, totals, medians, percentiles, ratios, and other statistics from survey data. It also provides univariate graphs such as histograms and boxplots of data from a complex survey.

The SURVEYFREQ procedure estimates percentages and performs categorical data analyses (such as chi-square tests) for data from complex surveys.

The SURVEYREG procedure performs linear regression analyses with data from a complex survey. It also creates multivariate displays of survey data.

The SURVEYLOGISTIC procedure performs logistic regression analyses with data from a complex survey.

DOI: 10.1201/9781003160366-1

SAS software has other procedures for performing analyses with survey data, such as the SURVEYPHREG procedure (which performs proportional hazard regression). These procedures have the same syntax as the ones explored in this book. Once you know how to use the SURVEYMEANS procedure, for example, it is easy to transfer that knowledge to the other procedures that analyze survey data.

Each of Chapters 2 through 12 contains code and output for the examples in that chapter, as well as tips for using SAS software and avoiding errors.

Conventions used in this book. SAS software does not distinguish between lowercase and uppercase characters, except for material between quotation marks. You can write code using all lowercase, ALL UPPERCASE, or a MiXTurE. In this book, I use uppercase to denote keywords and lowercase to denote the variable names and options that I supply. This is done solely so you can see which words are keywords and which words I supplied in the programs. The programs on the website are mostly in lowercase.

The names of external data files and programs, such as `agsrs.csv` and `example0205.sas`, are in `typewriter font`. Variable names and SAS data set names are in *italic type*.

Many of the examples in this book refer to figures, tables, examples, or exercises in *Sampling: Design and Analysis, Third Edition*, henceforth referred to as SDA. To avoid confusion, I refer to figures in SDA as "Figure x.x in SDA." I refer to figures in *this* book as "Figure x.x" with no qualifier.

All code displayed in this book is highlighted by a light gray shaded box and numbered consecutively within chapters. The output corresponding to a piece of code has the same number as the code: Output 2.1 contains output produced by Code 2.1. The caption for the code and output tells the name of the file, available on the companion website for the book (see the Preface for the website address), that contains the full set of code for that example. For example, the code printed in Code 2.1 and Code 2.2, which draws and prints a sample for Example 2.5 of SDA, is found in file `example0205.sas` on the website.

If you have used SAS software before. The remainder of this chapter introduces the basics of using SAS software: how to read data into the system, save data files, save output in various formats, and code missing data. If you have used SAS software before, you can skip the rest of this chapter and proceed to Chapter 2.

If you are new to SAS software. I suggest watching a couple of short video tutorials on reading data and doing basic calculations in the data step. You can find these at `https://video.sas.com/detail/videos/how-to-tutorials`. There are also numerous online tutorials and books on getting started with SAS software, where you can find much more detailed information than found in this brief chapter. A short introduction for those who have programmed in other languages, such as R, can be found at `https://support.sas.com/training/sas94/menunew.htm`.

I found it easiest to learn to use SAS software by example. When I need to write a program to analyze a new survey, I usually start by copying a program that does something similar and then modify the program for the current problem. If you are new to SAS software, the coding may seem complicated at first. But you will find as we move through the different types of survey data that once you learn how to use one of the procedures for survey data analysis, that knowledge carries over to all the other programs. The syntax is intuitive; all of the survey analysis procedures use the same basic set of commands, and they use them the same way.

1.1 Windows in SAS Software

When you start the program, you will see something that looks like Figure 1.1. Three windows are visible. The main windows referenced in this chapter are on the right 2/3 of the screen. On top is the log window, which gives error and warning messages. At the bottom is the editor window, in which you can type and edit programs.

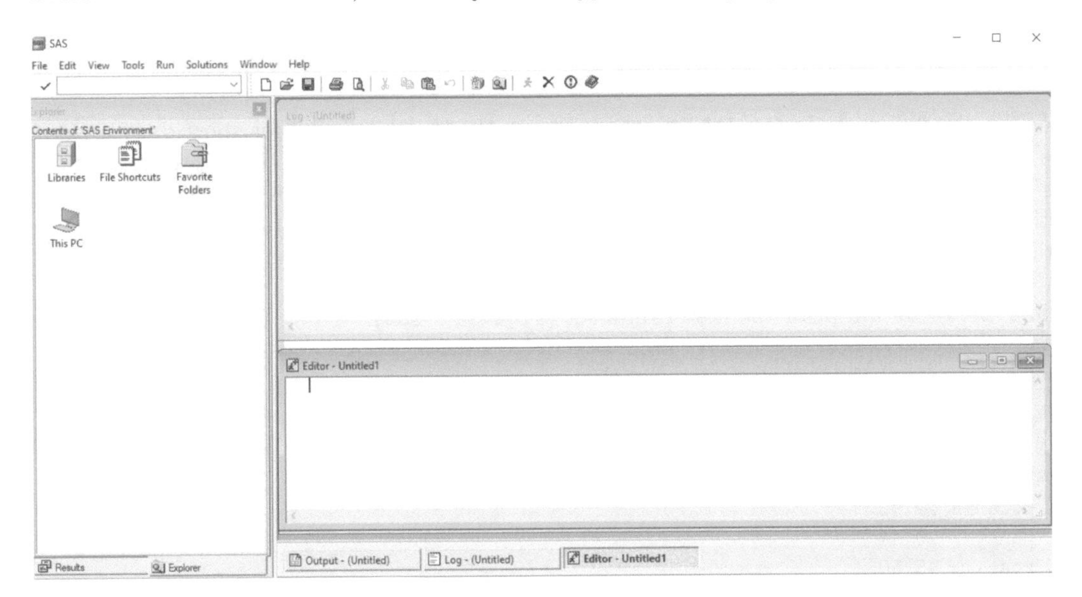

FIGURE 1.1: Editor and log window.

The toolbar at the top, when the Program Viewer is selected, contains icons for printing, saving files, and other standard activities. But it also contains four symbols specific for running SAS programs. These are shown and explained in Figure 1.2. The toolbar can be customized if desired.

Runs the selected SAS code.

Clear all. This erases the material in the selected window.

Break. This suspends processing of the submitted statements.

Help. This button opens the SAS help window, which gives more information on syntax for commands. You can also search online for help for specific commands.

FIGURE 1.2: Icons from the toolbar.

Getting help. There are lots of places to get help with SAS software. One source is the program itself: Click on the question-mark Help icon shown in Figure 1.2. You can also obtain online help and pdf manuals at `https://support.sas.com/en/documentation.html`. The manuals and help files contain additional examples of analyses (with code and annotated output) as well as the formulas used in computations. The *SAS/STAT® User's Guide* (SAS Institute Inc., 2021), available online, contains the full documentation for all SAS procedures used to select samples and analyze survey data.

1.2 Reading Data

The first step to using SAS software to analyze data from a survey is to read the data into the system. There are four basic ways to do this.

Enter the data directly in the DATA step. This method is best used for short data sets since you will be typing the data into the editor (program) window. Code 1.1 gives example code for entering a short data set containing the costs for three types of flowers and then printing it out. The code is also in file `flowers1.sas` on the book website (see the Preface for the website address).

Code 1.1. Read and print the flower data set (`flowers1.sas`).

```
/* Read the data values into data set mydata */

DATA mydata;
   INPUT id flower $ cost;
   DATALINES;
   1 petunia 29.95
   2 pansy   20.44
   3 rose    45.50
   ;
RUN;

/* Print the data in set mydata with title 'Flower Data'. */

PROC PRINT DATA=mydata;
   TITLE 'Flower Data';
RUN;
```

Let's walk through the key features of the DATA step in Code 1.1.

1. Semicolons are used to separate statements; you should put a semicolon at the end of every line. If the log window says there was an error in the code, first check whether you have forgotten to put a semicolon at the end of a line; this will often be the problem. It does not matter how many spaces you put between words or how many commands you put on one line. The indentation and spacing in the programs displayed in this book are my personal programming preferences; I have found these make the programs easier to read. You can use any format you feel comfortable with; just be sure to put semicolons in the right places.

2. Comments can be entered in two ways. Code 1.1 shows comments that start with slash-star (/*) and end with star-slash (*/). A line that starts with a star (*) also indicates a comment. Comments will help your future self remember what you did and why you did it; use them liberally.

3. The INPUT statement tells the variable names in the data set. Variables *id* and *cost* in this data set are numeric and are just listed in the statement. Variable *flower* is a character (text) variable. You tell the DATA step that a variable is of character type by adding a dollar sign ($) after the variable name. The data records, one per line, follow the DATALINES statement and the final semicolon marks the end of the data records to be read in.

4. The RUN statement tells the program to execute the code you just gave it.

5. As mentioned above, the DATA step and other procedures do not distinguish between uppercase and lowercase characters in variable names. The variable *FLOWER* is the same as the variable *flower*. You can type in all uppercase or all lowercase if you prefer.

You can either open the file `flowers1.sas` or type the commands into the 'Editor' window. After entering the code, run the program by clicking on the running figure (see Figure 1.2) in the toolbar (if you want to run just part of the code instead of all of it, select that part and then click on the running figure). Figures 1.3 and 1.4 show the windows after this program is run (click on the window name to switch from one window to another).

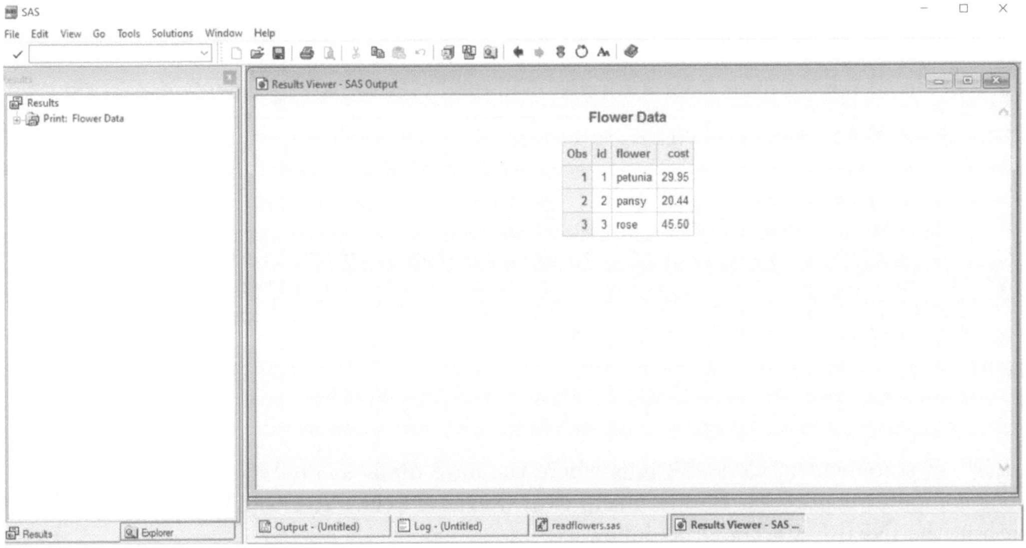

FIGURE 1.3: Results Viewer after running program `flowers1.sas`.

Figure 1.3 shows the output of the `PRINT` statement. It is tempting, once you have the output, to just paste it into your document and end the program. But you should always check the log window, shown for this example in Figure 1.4, before doing so. The log window tells you when you've made an error (these appear with the word ERROR in capital red letters) and provides other diagnostic information. There are no errors in Figure 1.4, so all is good. As we work through examples, I will show you some of the error and warning messages that appear.

Read data from a comma-delimited (.csv) or text file. With a longer data set, it is often more convenient to store the data in an external file and then read it in through the data step. Code 1.2 shows how to do that for the short flowers data set.

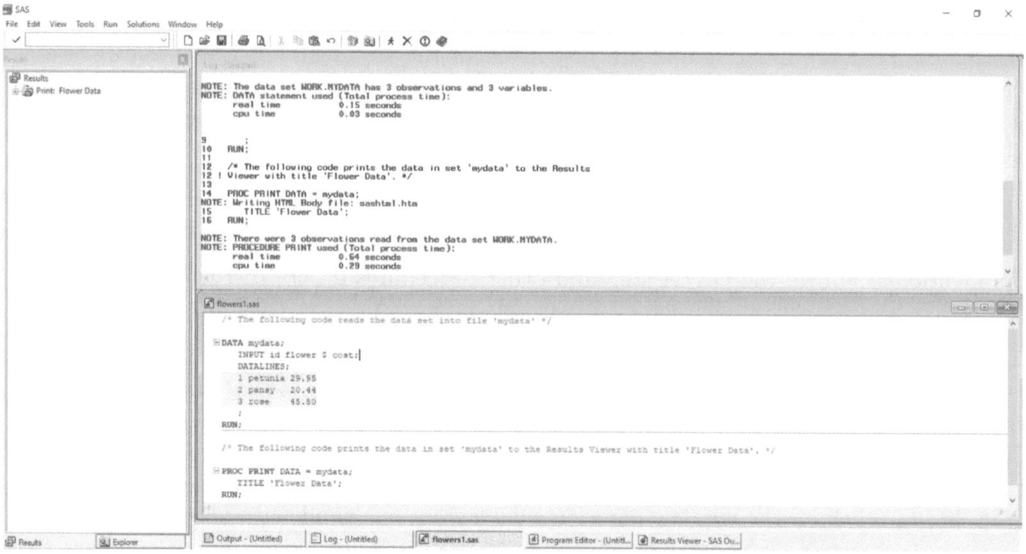

FIGURE 1.4: Editor and log windows after running program `flowers1.sas`.

Code 1.2. Read flowers data from a comma-delimited (`.csv`) file (`flowers2.sas`).

```
FILENAME flowdat "C:\MyFilePath\flowers.csv";

DATA mydata;
   INFILE flowdat DSD DELIMITER= ','  FIRSTOBS = 2;
     /* DELIMITER = ',' says this is a comma-delimited (.csv) file. */
     /* DSD option reads missing values between successive delimiters */
     /* The file flowers.csv has the variable names in the first line; FIRSTOBS=2
    says to skip the first line and start reading with line 2. */
   INPUT id flower $ cost;
RUN;
```

The INFILE statement in Code 1.2 tells where the data file is located. Suppose the folder on your computer containing the data sets is `C:\MyFilePath\`. You could specify the full filepath in the INFILE statement as follows:

```
INFILE "C:\MyFilePath\flowers.csv" DSD DELIMITER= ','  FIRSTOBS = 2;
```

but if you move the files to a different location later, you will then need to locate the INFILE statement in the middle of your program and change the filename. The FILENAME statement in the first line of Code 1.2 associates the name *flowdat* (note that the name is eight characters or less) with the file `C:\MyFilePath\flowers.csv`. Then, if you later move the data sets to a different directory, you need to change only the first line of the program.

Import data from a file. The command `Import Data` from the `File` menu will import the data from any format for you (comma-delimited, spreadsheet, text, tab-delimited, and others) and will also create the code for importing the data, which you can save for future use. The code created by `Import Data` is shown in Code 1.3 and also given in file `flowers3.sas`. Note that with the IMPORT procedure, you do not need to specify the variable names or types. The IMPORT procedure assigns the names in the first line to the variables, and uses the first few lines of the file to detect which variables are numeric and which are character.

In Code 1.3, the data in external file `C:\MyFilePath\flowers.csv` is read into the SAS data set *mydata*. The DBMS option specifies which type of file is read (here, a comma-delimited `.csv` file), and the REPLACE option says to replace the contents of data set *mydata* if it already exists.

Code 1.3. Import data from the external file `flowers.csv` (`flowers3.sas`).

```
PROC IMPORT OUT = work.mydata
            DATAFILE = "C:\MyFilePath\flowers.csv"
            DBMS = csv REPLACE;
     GETNAMES = yes;
     DATAROW = 2;
RUN;
```

After you run Code 1.3, the log window confirms that the observations in `flowers.csv` have been read into data set *mydata*.

```
NOTE: WORK.MYDATA data set was successfully created.
NOTE: The data set WORK.MYDATA has 3 observations and 3 variables.
NOTE: PROCEDURE IMPORT used (Total process time):
      real time           0.78 seconds
      cpu time            0.14 seconds
```

Read files that have been saved as SAS data sets. This is the easiest method of all—provided someone else has already saved the file as a SAS data set. All data files used in SDA have been saved as SAS data sets on the book's companion website. These have the suffix "`sas7bdat`"; for example, the filename for the flowers data is `flowers.sas7bdat`. The SAS data sets already contain the variable names and formatting information, so you can start running your analysis as soon as you read them in.

Code 1.4 and file `flowers4.sas` contain code to read in the data from `flowers.sas7bdat` and store it in data set *mydata*. The LIBNAME statement tells the directory name where the data file is located; here, I assigned the name *datalib* (you can call the library any name that has 8 or fewer characters) to the directory `C:\MyDataLib\`, where MyDataLib represents the path of the directory that contains the file `flowers.sas7bdat`. Then, any time you want to access data from that library, prefix the data set name with *datalib*.

Code 1.4. Read the SAS data set `flowers.sas7bdat` (`flowers4.sas`).

```
LIBNAME datalib "C:\MyDataLib\";

DATA mydata;
   SET datalib.flowers;
RUN;
```

1.3 Saving Output

Section 1.2 gave four methods for reading data into SAS software. How do you save the output from the program? Here are four methods that will allow you to save the output or to paste it into another document.

Copy and paste from the Results Viewer. As you work with SAS software, the output from procedures is displayed in the Results Viewer, as in Figure 1.3. You can save a specific bit of output by copying the selection and saving it to another document. This works quite well if you just want to save one graph or table from your output, but is inefficient if you want the output from the entire program.

Save the output from the Results Viewer to a file. You can save the entire output to a file by clicking on the Results Viewer window (to make it the active window), then selecting `Save As` from the `File` menu. Then specify a location and filename and choose one of the file types. You can save the output as a web archive, as a web page, or as a text file. If you have graphs in your output, choose the "Webpage, complete" file type. This will create a web page with your output and a folder that contains each graph as a separate `.png` image file.

Use ODS OUTPUT commands. You can save parts or all of your output in different formats through using the SAS Output Delivery System (ODS) commands. If you want to paste your output into a word-processing program, use the commands in Code 1.5 to save the output as an `.rtf` file. After running Code 1.5, the file `flowers.rtf` contains the output from the PRINT statement; in general, the file will contain the output for everything between the ODS RTF FILE= statement and the ODS RTF CLOSE statement.

Code 1.5. Direct output to an .rtf (rich text format) file (`flowers4.sas`).

```
FILENAME rtfout "C:\MyFilePath\flowers.rtf";

ODS RTF FILE=rtfout;
PROC PRINT data=mydata;
   TITLE 'Flower Data';
RUN;
ODS RTF CLOSE;
```

If you want to save your output in a different type of file, just substitute that file type in the ODS statement. I directed all the output for this book to pdf files by using code similar to that in Code 1.6.

Code 1.6. Direct output to a `.pdf` file (`flowers4.sas`).

```
FILENAME pdfout "C:\MyFilePath\Figure0101.pdf";

ODS PDF FILE=pdfout; /* Pdf file format */
/* Include code to create output or graph here */
ODS PDF CLOSE;
```

Run the program in BATCH mode. Once you have everything in your program working, you may want to run it all at once and save all the output in one place. The way to run a program in batch varies across operating systems. In Microsoft Windows, right-click on the program you want to run, then select "BATCH submit with SAS." If you do that with the program `flowers1.sas` (Code 1.1), two files will be created: `flowers1.log` contains the full log file, and `flowers1.lst` contains the text output, in typewriter font. If you have created any graphs, these will be saved in separate `.png` files.

1.4 Saving Data Sets

While working in SAS software, you may have created or modified data sets that you want to save. You can save these as SAS data sets, as spreadsheets or comma-delimited files, or in another format. Here's how.

Save a SAS data set. This is the easiest option. Let's say you want to save data set *mydata* as a SAS data set named `flowers2`. Then specify a directory where you want the data set to be saved, and simply save it to that location in a DATA step. Code 1.7 saves the file as `C:\MyFilePath\sasdata\flowers2.sas7bdat`.

Code 1.7. Save as a SAS data set (`flowers4.sas`).

```
LIBNAME datalib "C:\MyFilePath\sasdata"; /*Give the location of your SAS library.*/

DATA datalib.flowers2;
   SET mydata;
RUN;
```

The log window verifies that the program saved the file `flowers2.sas7bdat` in the directory path given in the LIBNAME statement:

```
NOTE: There were 3 observations read from the data set WORK.MYDATA.
NOTE: The data set DATALIB.FLOWERS2 has 3 observations and 3 variables.
NOTE: DATA statement used (Total process time):
      real time           0.01 seconds
      cpu time            0.01 seconds
```

Export the data set to a file. The EXPORT procedure saves the data set as an external file in the location desired. The command `Export Data` from the `File` menu will export the data in the format of your choice (.csv, spreadsheet, text, tab-delimited, and others). Or you can write the export instructions in code, as shown in Code 1.8 for writing the data set *mydata* to a comma-delimited file called `flowers2.csv`.

Code 1.8. Export the data set to a file (`flowers4.sas`).

```
PROC EXPORT DATA=mydata
   OUTFILE ='C:\MyFilePath\flowers2.csv'
   DBMS=csv
   REPLACE;
RUN;
```

1.5 Missing Data

Many survey data sets have observations that are missing. The survey data analysis procedures in SAS software have defaults for how missing data are treated (see Section 8.1 for details). For example, the SURVEYMEANS procedure, which we shall meet in Chapter 2, excludes observations with missing values from the analysis. But you must code the missing data so that the procedure recognizes that a value is missing.

The data file `agpop.csv`, from which the sample used in Chapter 2 is drawn, has missing data. As is common in data sets intended to be readable by multiple programs, a designated number is used to indicate that the data value is missing. In this data set, the value "−99" indicates that the data value is missing. This value must be recoded to the SAS symbol that indicates missing data before performing calculations. Otherwise, if, say, you want to calculate the mean of a variable, the procedure will treat all the "−99"s as if they were observations with the value −99 instead of missing values—this could lead to embarrassing results such as computing a negative value for the average number of acres per farm.

In SAS software, a single dot '.' denotes a missing value for a numeric variable, and an empty space ' ' denotes a missing value for a character variable. Code 1.9 reads the comma-delimited data in `agpop.csv`, recodes the missing observations to '.', and stores the modified data set as a SAS data set. The SAS data set in `agpop.sas7bdat` then has the correct missing value symbol for observations. Variables *county* and *state* are character (text) variables, so their names are followed by a $ sign in the INPUT statement. Because the names contained in variable *county* are long, I also add a LENGTH statement so that the county names will not be truncated to the standard length for a character variable.

Code 1.9. Recode missing data for *agpop* (`readagpop.sas`).

```
FILENAME agpop 'C:\MyFilePath\agpop.csv';
LIBNAME datalib 'C:\MyDataLib\';

DATA agpop;
   LENGTH county $ 26;  /* defines variable county to have length 26 */
   INFILE agpop DSD DELIMITER = ','  FIRSTOBS = 2;
   INPUT county $ state $ acres92 acres87 acres82 farms92 farms87 farms82 largef92
    largef87 largef82 smallf92 smallf87 smallf82 region $;
   /* recode the missing values */
   IF acres92 = -99 THEN acres92 =  . ;
   IF acres87 = -99 THEN acres87 =  . ;

   /* continue recoding all the other variables here */

   IF smallf87 = -99 THEN smallf87 =  . ;
   IF smallf82 = -99 THEN smallf82 =  . ;
RUN;

/* Save as a SAS data set */

DATA datalib.agpop;
   SET agpop;
RUN;
```

The missing data in the SAS data sets on the website have been recoded to the SAS missing data codes.

1.6 Summary, Tips, and Warnings

SAS software provides a powerful platform for selecting probability samples and analyzing data from surveys. A free cloud-based version of the software can be accessed from `https://www.sas.com/en_us/software/on-demand-for-academics.html`.

Two of the first tasks with any statistical software system are figuring out how to read data into it and how to save output from it. There are several ways to read data: typing directly into the program, reading in a text or comma-delimited file, or directly importing a spreadsheet or SAS data file. This chapter also outlines multiple ways of saving data sets and output that have been created in SAS software.

Tips and Warnings

- Always examine the log window after running a program. The log sometimes displays information about issues, warnings, and errors that are not apparent from the output in the Results Viewer.
- When the words 'Syntax Error' appear in the log, check to see if you forgot to put a semicolon at the end of a command (usually, the command before the line where the log locates the error). That often fixes the problem.
- Check the codes used for missing data in your data set before starting an analysis. Many public-domain data files use numbers (often negative numbers) to represent missing data. Recode these to '.' (or ' ' for a character variable) before starting your analysis.

2

Simple Random Sampling

Data from a simple random sample (SRS) can be analyzed using any statistical software designed for data that can be considered as independent and identically distributed. Sample means, sample variances, and sample proportions for an SRS can be estimated using the formulas taught in introductory statistics classes, and procedures such as the MEANS and FREQ procedures, intended for independent observations, correctly calculate these summary statistics.

To calculate statistics in SAS software from samples that are not SRSs, however, you need to use one of the survey analysis procedures. This chapter introduces you to the SURVEYMEANS and SURVEYFREQ procedures so that you start to become familiar with them, and shows how to use them to analyze data from an SRS. Subsequent chapters will show how to use these procedures with complex sample designs.

All data sets and code are available from the book website (see the Preface for the website address). The variables in the data sets are described in Appendix A.

Before calculating statistics, though, let's first look at how to use the SURVEYSELECT procedure to select an SRS from a population.

2.1 Selecting a Simple Random Sample

The SURVEYSELECT procedure will select almost any kind of probability sample for you. You need to supply it with a data set that lists the population units in the sampling frame, instructions for which type of sample to be drawn, and the name of the data set that is to contain the sample that is drawn. Code 2.1 shows how to do this for a small population of 10 units, with *id* numbers 1–10. SDA used a random number table to select an SRS of size 4 from this population.

Example 2.5 of SDA. *Selecting an SRS from a population.* The first step is to define a data set that contains identifiers for the N population units. You can read such a data set in from an external file, or generate it in a DATA step as done in Code 2.1. Here, a DO loop outputs the values 1 through 10 for variable *id* in the data set *pop*. The SURVEYSELECT procedure then draws an SRS of size 4 from data set *pop* and stores the sample in data set *srs4*. The full code for this example is also found in file `example0205.sas` on the companion website for this book.

DOI: 10.1201/9781003160366-2

Code 2.1. Select an SRS of size 4 from a population of size 10 (example0205.sas).

```
DATA pop;
   DO id = 1 TO 10;
      OUTPUT;
   END;  /* Always close a DO loop with END; */
RUN;

PROC SURVEYSELECT DATA=pop METHOD=SRS SAMPSIZE=4 OUT=srs4 STATS SEED=982736;
   TITLE 'Use the SURVEYSELECT procedure to draw SRS of size 4';
RUN;
```

Let's look at the form of the SURVEYSELECT procedure in Code 2.1, phrase by phrase:

PROC SURVEYSELECT tells which procedure to use. The SURVEYSELECT procedure, no surprise, selects a sample.

DATA= specifies the data set name (in this case, *pop*) that contains the population. For this example, *pop* lists the *id* numbers for the population from 1 to 10.

METHOD= specifies which method to use to draw the sample. Use METHOD=SRS to draw a simple random sample. Chapter 6 will discuss methods for drawing unequal probability samples and cluster samples.

SAMPSIZE= tells the sample size (n) of the sample to be drawn. Code 2.1 specifies SAMPSIZE=4.

OUT= gives the name of the data set containing the randomly selected sample. I called it *srs4*.

STATS requests that the selection probability and weight for each sampled unit be placed in the output data set. For an SRS, the selection probability is n/N for every sampled unit, and the weight is N/n, so this option could be omitted if you are willing to define the weight later.

SEED= specifies a number to use as a starting seed for the random number generation (I chose the seed 982736, but you can use any positive integer). If you omit the SEED option, a starting seed will be chosen using the internal clock. I always specify a seed explicitly, however, if I want to be able to reproduce the sample later. When you run the SURVEYSELECT procedure again with the same seed, you will get the same sample. If you run the same SURVEYSELECT procedure code again with a different seed, or with the SEED option omitted, you will get a different sample.

TITLE The TITLE statement is optional, but it helps you associate output with the code that generated it. You can enclose the title in single quotation marks ('My title') or double quotation marks ("My title"); use the latter if your title contains an apostrophe (for example, "Tom's output").

Now let's look at the output for this code in Output 2.1. The TITLE "Use the SURVEYSELECT procedure to draw SRS of size 4" is at the top of the output. The SURVEYSELECT procedure repeats the information you gave it for selecting the sample, and confirms that it drew an SRS from input data set *pop* using starting seed 982736. Each of the 4 observations in the SRS has selection probability 0.4 and sampling weight 2.5, and the sample is stored in the data set named *srs4*.

Output 2.1. Select an SRS of size 4 from a population of size 10 (`example0205.sas`).

Use the SURVEYSELECT procedure to draw SRS of size 4

The SURVEYSELECT Procedure

Selection Method	Simple Random Sampling

Input Data Set	POP
Random Number Seed	982736
Sample Size	4
Selection Probability	0.4
Sampling Weight	2.5
Output Data Set	SRS4

Because this is a small sample, let's print the whole thing using Code 2.2. For a larger data set, you may want to print just a few observations (use the OBS= option of the PRINT statement to do this; see Code 2.3).

Code 2.2. Print the SRS of size 4 (`example0205.sas`).

```
PROC PRINT DATA = srs4;
   TITLE 'Print data set srs4';
RUN;
```

Output 2.2. Print the SRS of size 4 (`example0205.sas`).

Print data set srs4

Obs	id	SelectionProb	SamplingWeight
1	1	0.4	2.5
2	4	0.4	2.5
3	5	0.4	2.5
4	10	0.4	2.5

The log confirms that an SRS of 4 observations was selected and that 4 observations of the data set *srs4* were printed. I shall not display logs in this book for subsequent examples of the SURVEYSELECT procedure except to point out warnings, but you should always check the log for errors after running a procedure.

```
NOTE: The data set WORK.SRS4 has 4 observations and 3 variables.
NOTE: PROCEDURE SURVEYSELECT used (Total process time):
      real time           0.10 seconds
      cpu time            0.03 seconds

NOTE: There were 4 observations read from the data set WORK.SRS4.
NOTE: PROCEDURE PRINT used (Total process time):
      real time           0.01 seconds
      cpu time            0.01 seconds
```

Example 2.6 of SDA. The sample in `agsrs.csv` was selected from the population using random numbers generated in a spreadsheet. Code 2.3 shows code that could be used to select another SRS of size 300 from the 3078 counties in file `agpop.csv`. This sample will differ from the one in SDA because a different procedure is used. I print the first 10 observations to check that the sample was drawn and to see the variable names created by the SURVEYMEANS procedure.

Code 2.3. Select an SRS from the data in `agpop.csv` (`example0206.sas`).

```
LIBNAME datalib 'C:\MyDataLib\';

DATA agpop;
   SET datalib.agpop;
RUN;

PROC SURVEYSELECT DATA=agpop METHOD = srs SAMPSIZE = 300 OUT = agsrs2 STATS SEED =
     88837264;
   TITLE 'Draw SRS of size 300';
RUN;

PROC PRINT DATA=agsrs2 (OBS = 10);
   TITLE 'Print first 10 observations from agsrs2';
RUN;
```

Output 2.3(a) shows the output from the SURVEYSELECT procedure. As seen in the printout of the first 10 observations in Output 2.3(b), the procedure puts the SRS selection probabilities in variable *SelectionProb* and the sampling weights in variable *SamplingWeight.*

Output 2.3(a). Select an SRS from the data in `agpop.csv` (`example0206.sas`).

Draw SRS of size 300

The SURVEYSELECT Procedure

Selection Method	Simple Random Sampling

Input Data Set	AGPOP
Random Number Seed	88837264
Sample Size	300
Selection Probability	0.097466
Sampling Weight	10.26
Output Data Set	AGSRS2

Output 2.3(b). Print the first 10 observations in the sample *agsrs2* (`example0206.sas`).

Print first 10 observations from agsrs2

Obs	county	state	acres92	acres87	acres82	farms92	farms87	farms82	largef92
1	ALEUTIAN ISLANDS AREA	AK	683533	726596	764514	26	27	28	14
2	FAIRBANKS AREA	AK	141338	154913	204568	168	175	170	25
3	CLARKE COUNTY	AL	61426	94506	116008	213	331	429	13
4	CRENSHAW COUNTY	AL	111315	118184	134393	458	494	558	21
5	ELMORE COUNTY	AL	104364	128572	141703	519	622	713	15
6	ETOWAH COUNTY	AL	85821	100517	117624	774	928	998	6
7	JACKSON COUNTY	AL	204487	208014	226697	1139	1224	1295	33
8	LAWRENCE COUNTY	AL	173468	188365	201504	915	1123	1226	27
9	TALLADEGA COUNTY	AL	104199	114668	128253	472	552	630	18
10	HOWARD COUNTY	AR	105721	102407	111503	658	656	674	13

Obs	largef87	largef82	smallf92	smallf87	smallf82	region	SelectionProb	SamplingWeight
1	16	20	6	4	1	W	0.097466	10.26
2	28	21	12	18	25	W	0.097466	10.26
3	15	21	10	14	20	S	0.097466	10.26
4	22	21	26	33	28	S	0.097466	10.26
5	19	25	19	31	36	S	0.097466	10.26
6	10	12	48	61	56	S	0.097466	10.26
7	33	33	40	60	83	S	0.097466	10.26
8	41	34	42	59	85	S	0.097466	10.26
9	18	22	13	25	31	S	0.097466	10.26
10	12	11	42	54	39	S	0.097466	10.26

2.2 Computing Statistics from an SRS

All of the statistics discussed in Chapter 2 of SDA can be computed in SAS software using the SURVEYMEANS procedure. The keywords used to request the statistics, along with standard errors and confidence intervals, are listed in Table 2.1. The SURVEYMEANS procedure also calculates many other statistics; for the full list, see the help file or documentation for the procedure (SAS Institute Inc., 2021).

Numeric and categorical variables. Table 2.1 mentions calculations for numeric and categorical variables. In SAS software, numeric variables are variables for which you want to calculate statistics such as means. Categorical variables are those for which the values represent categories.

A variable *height*, recording the height for each person in a data set, is numeric. We may want to calculate the mean or median height for the sample values. A variable *haircolor*, with categories *black*, *brown*, *blond*, *red*, *bald*, and *other*, is categorical. We may want to estimate the proportion of persons in the population who are in each category, but we cannot calculate an average hair color.

Some surveys code categories as numbers; be careful to treat such variables as categorical rather than numeric. For example, the variable *hair* might take on values 1–6, where 1 represents black, 2 represents brown, 3 represents blond, 4 represents red, 5 represents bald,

TABLE 2.1
Statistics calculated by the SURVEYMEANS procedure, used in Chapter 2 of SDA.

Keyword	Statistic
ALL	requests all statistics for the variable
CLM	confidence interval for the MEAN
CLSUM	confidence interval for the population total (SUM)
CV	coefficient of variation for MEAN
CVSUM	coefficient of variation for SUM
DF	degrees of freedom (df) used for t tests and t confidence intervals
MEAN	mean (if the variable is numeric), or proportion in each category (if variable is categorical)
MEDIAN	median (50th percentile) if variable is numeric
NMISS	number of observations with missing data
NOBS	number of observations with data
QUARTILES	25th percentile, 50th percentile (MEDIAN), and 75th percentile for a numeric variable
STD	standard error for estimated population total (SUM)
STDERR	standard error for estimated MEAN
SUM	estimated population total, using the weights, $\sum_{i \in \mathcal{S}} w_i y_i$
SUMWGT	sum of the weights, $\sum_{i \in \mathcal{S}} w_i$
VAR	estimated variance for MEAN
VARSUM	estimated variance for SUM

and 6 represents other. You can calculate the average of the numbers in the variable *hair*, but it is meaningless.

In the SURVEYMEANS procedure, you specify that a variable such as *hair* is categorical by using the CLASS statement. If you omit the CLASS statement, the SURVEYMEANS procedure will treat the variable as numeric.

Examples 2.6, 2.7, 2.8, 2.9, and 2.11 of SDA. We can obtain all the statistics displayed in these examples, and draw a histogram of the data, using the SURVEYMEANS procedure. The DATA step in Code 2.4 (the code is also in file `example0211.sas` on the book website) reads the SAS data file `C:\MyDataLib\agsrs.sas7bdat` into data set *agsrs*. It then calculates the variable *lt200k*, which takes on the value 1 if the county has less than 200,000 acres in farms and takes on the value 0 if the county has 200,000 or more acres in farms. The data set *agsrs* does not contain a variable with the sampling weights, so the weight variable *sampwt* is defined in the DATA step to equal 3078/300 for each observation in the sample.

Code 2.4. Define new variables for *agsrs* (`example0211.sas`).

```
LIBNAME datalib 'C:\MyDataLib\';

/* Read data set and calculate new variable lt200k.
   Also define sampwt to be the sampling weight for each observation. */

DATA agsrs;
   SET datalib.agsrs;
   IF acres92 < 200000 THEN lt200k = 1; /* counties with < 200000 acres in farms */
   ELSE IF acres92 >= 200000 THEN lt200k = 0;
   sampwt = 3078/300;  /* sampling weight is same for each observation */
RUN;
```

Code 2.5. Calculate estimates from *agsrs* with the SURVEYMEANS procedure (`example0211.sas`).

```
PROC SURVEYMEANS DATA = agsrs TOTAL=3078 PLOTS=ALL SUM CLSUM MEAN CLM CV MEDIAN;
   CLASS lt200k;
   WEIGHT sampwt;
   VAR acres92 lt200k;
   TITLE 'Analysis of SRS from Census of Agriculture';
RUN;
```

The statistics in Examples 2.6, 2.7, 2.8, 2.9, and 2.11 of SDA are calculated using Code 2.5. Let's look at the commands in the SURVEYMEANS procedure in Code 2.5, phrase by phrase:

PROC SURVEYMEANS requests the SURVEYMEANS procedure.

DATA= specifies the data set (in this case, *agsrs*) to be analyzed.

TOTAL= tells the size of the population (N) from which the sample was drawn. For these data, $N = 3078$. This gives the procedure the information it needs to calculate a finite population correction (fpc). If you omit the TOTAL= phrase, estimates will be computed without the fpc.

PLOTS= ALL requests all available plots. For this example, a histogram and boxplot are drawn for each numeric variable specified in the VAR statement. If you wish to suppress the plots, type `PLOTS = NONE`.

SUM CLSUM MEAN CLM CV MEDIAN lists the keywords for the set of statistics to be calculated (see Table 2.1). If in doubt about which statistics you will need, just type ALL.

CLASS *class_var1 ... class_varp* declares the variables *class_var1 ... class_varp* to be categorical. Code 2.5 declares *lt200k* to be a categorical variable in the CLASS statement. If you omit the CLASS statement, all variables will be treated as numeric.

WEIGHT *weight_var* says to use variable *weight_var* as the weight variable. I used variable *sampwt*. Always use a weight statement in PROC SURVEYMEANS. If you omit it, all weights are assumed to be 1. For an SRS, estimated means will be correct if weights are assumed to be 1, but estimated population totals will be wrong. Even if the weights really are all equal to 1, it is good practice to define the weight variable explicitly.

VAR *var1 var2 ... vark class_var1 ... class_varp* tells the program to calculate statistics for numeric variables *var1 var2 ... vark* and categorical variables *class_var1 ... class_varp* (make sure you define the categorical variables as such in a CLASS statement). In Code 2.5, statistics for *acres92* and *lt200k* are requested. You can put as many variables as desired following the VAR keyword.

After running the program, check the log file for errors. If there are none, the log will appear as follows (this is how it appears for most programs where no errors are found; for subsequent programs in this book, I will include the log file only if it shows an error or warning):

```
NOTE: PROCEDURE SURVEYMEANS used (Total process time):
      real time           1.22 seconds
      cpu time            0.68 seconds
```

Let's first look at the histogram and boxplot produced for the numeric variable *acres92* by the PLOTS=ALL option. The plots produced by the SURVEYMEANS procedure are discussed in more generality in Section 7.4 of this book and in Chapter 7 of SDA. You can customize these plots by selecting which plots to display, how many bins to use in the histogram, and other options.

The histogram in Output 2.5(a) is accompanied by two density estimates. The solid line shows the density function of a normal distribution with mean and standard deviation equal to the estimated mean and standard deviation of variable *acres92*. The normal distribution does not follow the shape of the data well; the kernel density estimate, described in Section 7.6 of SDA, is a smoothed version of the histogram and shows that the data have a skewed distribution.

The horizontal shaded box in the boxplot shows the estimated median and quartiles of *acres92*, and the darker gray vertical box that crosses the boxplot displays the 95% confidence interval for the mean. You can also use the SGPLOT procedure to produce customized plots for survey data, as will be seen in Section 7.4.

Output 2.5(a). Histogram and boxplot for *acres92* from the SURVEYMEANS procedure (`example0211.sas`).

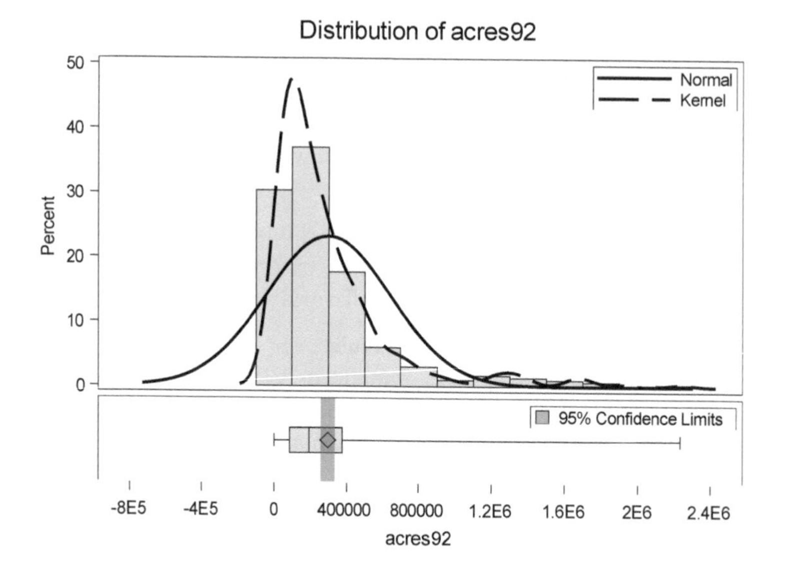

Now let's look at the statistics produced by the SURVEYMEANS procedure, shown in Output 2.5(b). The first box shows a "Data Summary," which gives the number of observations in the data set and the sum of the weights. For an SRS, the weights sum to the population size (here, 3078).

The second box in Output 2.5(b) describes the levels taken on by variable *lt200k*, which was defined to be a CLASS (categorical) variable.

The third and fourth boxes in Output 2.5(b) give the statistics calculated by the SURVEYMEANS procedure. The statistics calculated for numeric variable *acres92* are, from left to right: sample mean $\bar{y}$, standard error of the mean SE $(\bar{y})$, 95% confidence interval for the mean, coefficient of variation of the mean [$=$ SE $(\bar{y})/\bar{y}$], estimated population total $\hat{t}$, standard error of the total SE $(\hat{t})$, and 95% confidence interval for the population total. The standard errors and confidence intervals are calculated using a finite population correction because the TOTAL is specified in the PROC SURVEYMEANS statement. For an SRS, the default is to calculate confidence intervals using a t distribution with $n-1$ degrees of freedom.[1] The statistics in the output are the same as those in Examples 2.7 and 2.11 of SDA.

The "Quantiles" box gives the estimate and 95% confidence limits for the median (standard errors and confidence intervals for medians are discussed in Chapter 7 of SDA). The SURVEYMEANS procedure gives a slightly different estimate of the median (196,701) than the estimate given in Example 2.7 of SDA (196,717) because it uses a slightly different formula for calculating the median (this will be discussed in Section 7.2), but both formulas give consistent estimators of the population median when the variable's distribution is approximately continuous.

Output 2.5(b). Calculate estimates from *agsrs* with the SURVEYMEANS procedure (`example0211.sas`).

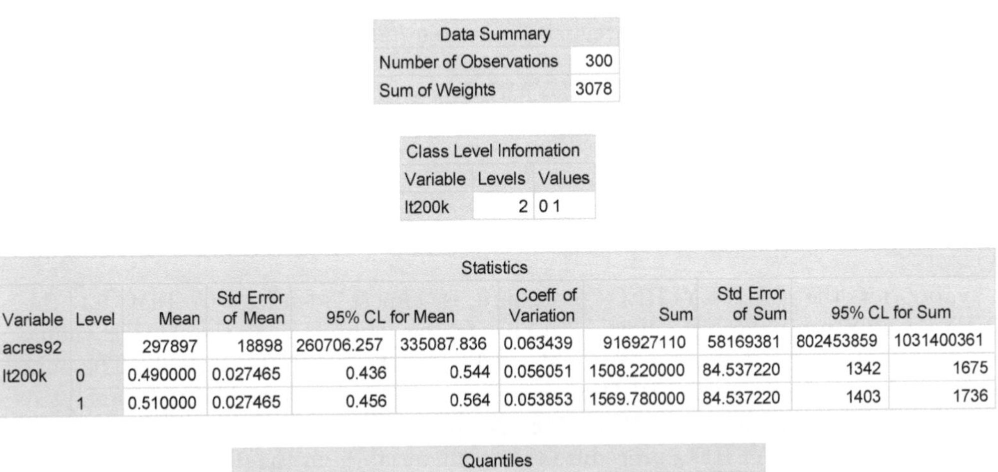

Analysis of SRS from Census of Agriculture

The SURVEYMEANS Procedure

Data Summary	
Number of Observations	300
Sum of Weights	3078

Class Level Information		
Variable	Levels	Values
lt200k	2	0 1

Statistics

Variable	Level	Mean	Std Error of Mean	95% CL for Mean		Coeff of Variation	Sum	Std Error of Sum	95% CL for Sum	
acres92		297897	18898	260706.257	335087.836	0.063439	916927110	58169381	802453859	1031400361
lt200k	0	0.490000	0.027465	0.436	0.544	0.056051	1508.220000	84.537220	1342	1675
	1	0.510000	0.027465	0.456	0.564	0.053853	1569.780000	84.537220	1403	1736

Quantiles

Variable	Percentile		Estimate	Std Error	95% Confidence Limits	
acres92	50	Median	196701	19529	158268.707	235133.293

For categorical variable *lt200k*, the SURVEYMEANS procedure calculates the proportion in each category, and the standard errors and 95% confidence intervals for the category proportions. The SUM column gives the estimated number of population members in each category, along with standard error and 95% confidence interval. The value of $\hat{p} = 0.51$ from Example 2.8 in SDA is the estimated proportion where *lt200k* takes on the value 1.

[1]Most survey data sets are large enough that the degrees of freedom make essentially no difference for the confidence intervals. For this example, the t critical value with 299 degrees of freedom is 1.968; the corresponding critical value from a normal distribution is 1.960.

2.3 Estimating Proportions from an SRS

As we saw in Section 2.2, the SURVEYMEANS procedure will estimate proportions. For a binary numeric variable (taking on values 0 or 1), the estimated proportion is the mean. You can estimate proportions of different categories for a categorical variable in the SURVEYMEANS procedure by defining the variable to be categorical in a CLASS statement. For all examples and exercises in Chapter 2 of SDA except Exercise 2.34 (see Section 2.4), you can calculate estimated proportions using the SURVEYMEANS procedure.

The SURVEYFREQ procedure will also estimate proportions, and it has some additional options that are sometimes useful for survey data. For example, if the estimated proportion is close to 0 or 1, the SURVEYFREQ procedure can be used to compute an asymmetric confidence interval (see Section 2.4) instead of the symmetric confidence interval computed by the SURVEYMEANS procedure. In this section, though, let's look at how to estimate proportions with confidence intervals of the form $\hat{p} \pm t\,\mathrm{SE}\,(\hat{p})$, where t represents a critical value of the t distribution, and compare the output from the SURVEYFREQ procedure with that from the SURVEYMEANS procedure.

Examples 2.9 and 2.11 of SDA. Code 2.6 shows how to use the SURVEYFREQ procedure to estimate the percentages in each category of variable *lt200k*.

Code 2.6. Estimates from `agsrs`, SURVEYFREQ procedure (`example0211.sas`).

```
PROC SURVEYFREQ DATA = agsrs TOTAL = 3078;
   WEIGHT sampwt;
   TABLES lt200k / CL CLWT;
RUN;
```

The syntax for the SURVEYFREQ procedure is almost identical to that for the SURVEYMEANS procedure. For both, you specify the input DATA, the TOTAL number of units in the population (if an fpc is desired), and the WEIGHT variable. The only difference is that in the SURVEYFREQ procedure, you specify the variables to be analyzed in a TABLES statement (in the SURVEYMEANS procedure, you specify them in a VAR statement). The SURVEYFREQ procedure treats all variables in the TABLES statement as categorical, so no CLASS statement is used with the procedure. The keyword CL asks for confidence limits for the estimated proportion, and the keyword CLWT asks for confidence limits for the estimated population totals (weighted frequencies) in each category.

The statistics calculated in Code 2.6 are the same as those from the SURVEYMEANS procedure in Code 2.5; the only difference is that Output 2.6 presents percentages, which equal the proportions in Output 2.5(b) multiplied by 100. The weighted frequencies in Output 2.6 are the estimated number of population members in each class, and are the same as the statistics in the "Sum" column of Output 2.5(b).

Output 2.6. Estimates from *agsrs*, SURVEYFREQ procedure (`example0211.sas`).

The SURVEYFREQ Procedure

Data Summary	
Number of Observations	300
Sum of Weights	3078

Table of lt200k									
lt200k	Frequency	Weighted Frequency	Std Err of Wgt Freq	95% Confidence Limits for Wgt Freq		Percent	Std Err of Percent	95% Confidence Limits for Percent	
0	147	1508	84.53722	1342	1675	49.0000	2.7465	43.5951	54.4049
1	153	1570	84.53722	1403	1736	51.0000	2.7465	45.5951	56.4049
Total	300	3078	0	3078	3078	100.0000			

2.4 Additional Code for Exercises

Exercise 2.27 of SDA. Figure 2.8 of SDA contains a histogram of the means of 1,000 samples of size 300 taken with replacement from *agsrs*. Code 2.7 shows how you can draw such a histogram (using a different set of 1,000 samples than in SDA). In the SURVEYSELECT procedure, METHOD=URS requests simple random samples with replacement; REPS=1000 says to draw 1,000 independent samples.

Code 2.7. Drawing a histogram similar to that in Figure 2.8 of SDA (`exercise0227.sas`).

```
PROC SURVEYSELECT DATA = agsrs N = 300 METHOD = URS REPS = 1000 OUT = bootout SEED
    = 3422957;

/* Calculate the mean for each replicate sample, store in variable repmean */
/* The NOPRINT option suppresses the pages of output that would otherwise be
    generated. Instead, the OUTPUT statement stores the 1000 means in data set
    bootmeans. */
PROC MEANS DATA = bootout NOPRINT;
   BY replicate;
   VAR acres92;
   OUTPUT out = bootmeans MEAN = repmean;

PROC SGPLOT DATA=bootmeans;
   HISTOGRAM repmean;
   XAXIS label = "Means from 1000 with-replacement samples";
RUN;
```

Exercise 2.34 of SDA. Clopper-Pearson confidence intervals are often a better choice than symmetric intervals when estimated proportions are close to zero or one.

The SURVEYFREQ procedure will compute Clopper-Pearson and other types of confidence intervals for you. Code 2.8 shows you how. The only difference from the code in Code 2.6, which used the normal approximation to calculate confidence intervals, is the `(TYPE=clopperpearson)` option behind the CL request. Output 2.8 shows that for this example, the Clopper-Pearson intervals are the same as those based on the normal approximation. This is true because here, $\hat{p} = 0.51$ is not close to 0 or 1 and the sample size is relatively large. When the estimated proportion is close to 0 or 1, however, as seen in Exercise 2.34, the confidence intervals differ. A symmetric confidence interval of the form $\hat{p} \pm 1.96\,\text{SE}(\hat{p})$ can include values that are less than 0 or greater than 1, while the Clopper-Pearson interval is restricted to be in the interval [0, 1].

Code 2.8. Clopper-Pearson confidence intervals (`example0211.sas`).

```
PROC SURVEYFREQ DATA = agsrs TOTAL = 3078;
   WEIGHT sampwt;
   TABLES lt200k / CL(TYPE=clopperpearson) CLWT;
   TITLE 'Clopper-Pearson confidence intervals for variable lt200k';
RUN;
```

Output 2.8. Clopper-Pearson confidence intervals (`example0211.sas`).

Clopper-Pearson confidence intervals for variable lt200k

The SURVEYFREQ Procedure

Data Summary	
Number of Observations	300
Sum of Weights	3078

Table of lt200k

lt200k	Frequency	Weighted Frequency	Std Err of Wgt Freq	95% Confidence Limits for Wgt Freq		Percent	Std Err of Percent	95% Confidence Limits for Percent	
0	147	1508	84.53722	1342	1675	49.0000	2.7465	43.2109	54.8091
1	153	1570	84.53722	1403	1736	51.0000	2.7465	45.1909	56.7891
Total	300	3078	0	3078	3078	100.0000			

Clopper-Pearson (exact) confidence limits are computed for percents.

2.5 Summary, Tips, and Warnings

The general structure for selecting an SRS using the SURVEYSELECT procedure is:

```
PROC SURVEYSELECT DATA=pop_dataset METHOD = srs SAMPSIZE = desired_sample_size
    OUT = srs_dataset STATS SEED = seed_integer;
```

The STATS and SEED= phrases are optional for an SRS, but I usually include them to obtain the sampling weights and to be able to re-create the sample using the same seed later. If you omit SEED=, the procedure will generate a seed for you and print it in the output.

The general structure for the SURVEYMEANS procedure to analyze data from an SRS is:

```
PROC SURVEYMEANS DATA=srs_dataset TOTAL=pop_total statistics_keywords;
   CLASS class_var1 class_var2 class_var3;
   WEIGHT weight_var;    /* Always include a weight variable */
   VAR var1 var2 var3 class_var1 class_var2 class_var3;
```

I listed three numeric variables (*var1 var2 var3*) and three categorical variables (*class_ var1 class_ var2 class_ var3*) in the VAR statement, but you can include as many variables in the statement as you want to analyze. Table 2.1 lists some of the commonly requested statistics keywords for the SURVEYMEANS procedure.

The CLASS statement should be included if there are any categorical variables to be analyzed (here, *class_ var1 class_ var2 class_ var3*) in the data set. If there are no categorical variables, do not include a CLASS statement.

The general structure for the SURVEYFREQ procedure to analyze data from an SRS is:

```
PROC SURVEYFREQ DATA=srs_dataset TOTAL=pop_total;
   WEIGHT weight_variable;    /* Always include a weight variable */
   TABLES class_var1 class_var2 class_var3 / CL CLWT;
```

I listed three categorical variables (*class_ var1 class_ var2 class_ var3*) in the TABLES statement, but you can include as many variables as you want to analyze. No CLASS statement is needed because the SURVEYFREQ procedure assumes that all variables in the TABLES statement are categorical. Options CL and CLWT of the TABLES statement request confidence intervals for percentages and totals.

The phrase `TOTAL=pop_total` is optional in the SURVEYMEANS and SURVEYFREQ procedures; if used, `pop_total` is the population size N and an fpc is used in standard error calculations.

Tips and Warnings

- If you want to be able to draw the same sample from a population at a later date, use the SEED option in the SURVEYSELECT procedure.
- Always include a WEIGHT statement when analyzing survey data in the SURVEYMEANS or SURVEYFREQ procedure. If you omit this statement, a weight of 1 will be assigned to every observation.
- In the output for the SURVEYMEANS and SURVEYFREQ procedures, check that the sum of the weights is approximately equal to the population size. If not, you have made a mistake in computing weights or in specifying the weight variable in the program.
- If you are analyzing categorical variables in the SURVEYMEANS procedure, make sure to tell the program they are categorical in a CLASS statement. Otherwise, the procedure will treat them as numeric variables and calculate the mean instead of computing the proportion in each category.

3

Stratified Sampling

Stratified random samples have two features that require the use of survey data analysis programs such as the SURVEYMEANS and SURVEYFREQ procedures. Unequal weights from disproportionally allocated samples must be used in the estimation to obtain unbiased estimates of population means and totals. The stratification also affects the standard errors of estimates.

Sample selection, too, is more complicated than for a simple random sample. Fortunately, the SURVEYSELECT procedure makes it easy to draw a stratified random sample. You need only provide information about the strata and desired stratum sizes to obtain the type of allocation desired.

3.1 Selecting a Stratified Random Sample

The SURVEYSELECT procedure will calculate sample sizes for proportional and optimal allocations and select the sample. You can also specify a custom allocation. Let's start by drawing a proportionally allocated sample.

Example 3.2 of SDA. I used a spreadsheet to select a stratified random sample from the population in `agpop.csv`, but let's look at how to do that with the SURVEYSELECT procedure (this will, of course, give a different sample than obtained in Example 3.2 of SDA). Code 3.1 selects a stratified random sample of size 300 with proportional allocation. In this and subsequent code in the book, I assume that you have already read the data into the SAS data set (here, *agpop*) from the external file using one of the methods discussed in Section 1.2. I shall also omit all but the last RUN statement from code, but you should use a RUN statement every time you want a procedure to execute.

Code 3.1. Select a stratified random sample of size 300 from *agpop* with proportional allocation (`example0302.sas`).

```
PROC SORT DATA = agpop;
  BY region;

PROC SURVEYSELECT DATA = agpop METHOD = SRS SAMPSIZE = 300 OUT = agstrat2 STATS
    SEED = 82835264;
  STRATA region / ALLOC = prop;
  TITLE 'Select proportionally allocated stratified random sample of size 300';
RUN;
```

The code is similar to that used to select an SRS in Code 2.3. Let's look at the features that specify selecting a stratified sample.

DOI: 10.1201/9781003160366-3

PROC SORT sorts the data set by the stratification variable *region*. **Always** sort the population data set before selecting the stratified sample to avoid errors.

PROC SURVEYSELECT The statement is basically the same as in Code 2.3, specifying the data set name, the METHOD (=SRS since we want to take an SRS in each stratum), the data set name to contain the sample (OUT= *agstrat2*), and the random number SEED. The only part that is interpreted differently is the sample size SAMPSIZE.

SAMPSIZE= With stratified sampling, this is interpreted differently depending on whether you use the ALLOC= option in the STRATA statement. If you include the ALLOC= option, SAMPSIZE specifies the total sample size to be divided among all the strata, so in Code 3.1, SAMPSIZE=300 specifies that the total sample size is 300. If you omit the ALLOC= option and specify, say, SAMPSIZE=10, an SRS of size 10 is taken from each stratum, so that the total sample size is 10 × (number of strata).

STRATA region specifies that a stratified sample is to be drawn using stratification variable *region*. If you include more than one stratification variable, say by typing

```
STRATA stratvar1 stratvar2;
```

where *stratvar1* has 3 levels and *stratvar2* has 2 levels, then 6 strata are defined by the cross-classification of *stratvar1* and *stratvar2*.

You tell which type of allocation to use as an option in the STRATA statement, behind the slash (/). I wanted proportional allocation, so I wrote ALLOC=prop.

Output 3.1 repeats the specifications you provided for selecting the sample. Check that the total sample size and allocation method are what you intended.

Output 3.1. Select a stratified random sample of size 300 from *agpop* with proportional allocation (`example0302.sas`).

The SURVEYSELECT Procedure

Selection Method	Simple Random Sampling
Strata Variable	region
Allocation	Proportional

Input Data Set	AGPOP
Random Number Seed	82835264
Number of Strata	4
Total Sample Size	300
Output Data Set	AGSTRAT2

Now let's print the sample sizes in each stratum and the first few observations of the sample to check, in Code 3.2. I will not show these checks for the other samples drawn in this section, but it is good practice to run the FREQ procedure after drawing a sample just to make sure that the SURVEYSELECT procedure did what you wanted it to. The PRINT procedure shows some of the variables for the first six observations of the sample (in this case, all from the NC region because the file is sorted by region).

The SURVEYSELECT procedure creates variables *Total* (the population size N_h in the stratum), *AllocProportion* (= N_h/N for proportional allocation, the proportion to be allocated to the stratum), *SampleSize* (the sample size drawn from the stratum), and *ActualProportion* (= n_h/n, the actual proportion drawn from the stratum). It also creates variables containing the selection probability and the sampling weight for each observation.

For this example, *AllocProportion* differs slightly from *ActualProportion* because the sample sizes must be integers. For the NC stratum, *AllocProportion* $= 1054/3078 = 0.34243$ while *ActualProportion* $= 103/300 = 0.34333$.

Code 3.2. Check the results of the proportional allocation (`example0302.sas`).

```
PROC FREQ DATA = agstrat2;
   TABLES region;
   TITLE 'Check sample sizes in each stratum of proportionally allocated sample';

PROC PRINT DATA = agstrat2 (OBS = 6);
   VAR region acres92 Total AllocProportion SampleSize ActualProportion
    SelectionProb SamplingWeight;
   TITLE 'Print the first few observations of selected variables';
RUN;
```

Output 3.2. Check the results of the proportional allocation (`example0302.sas`).

Check sample sizes in each stratum of proportionally allocated sample

The FREQ Procedure

region	Frequency	Percent	Cumulative Frequency	Cumulative Percent
NC	103	34.33	103	34.33
NE	21	7.00	124	41.33
S	135	45.00	259	86.33
W	41	13.67	300	100.00

Print the first few observations of selected variables

Obs	region	acres92	Total	AllocProportion	SampleSize	ActualProportion	SelectionProb	SamplingWeight
1	NC	236668	1054	0.34243	103	0.34333	0.097723	10.2330
2	NC	315448	1054	0.34243	103	0.34333	0.097723	10.2330
3	NC	456954	1054	0.34243	103	0.34333	0.097723	10.2330
4	NC	192467	1054	0.34243	103	0.34333	0.097723	10.2330
5	NC	366927	1054	0.34243	103	0.34333	0.097723	10.2330
6	NC	328885	1054	0.34243	103	0.34333	0.097723	10.2330

3.1.1 Allocation Methods

Proportional allocation. Code 3.1 specifies proportional allocation. Write ALLOC=prop or ALLOC=proportional behind the slash in the STRATA statement. In the SURVEYSELECT statement, set the SAMPSIZE= argument equal to the total sample size to be divided among the strata.

Neyman allocation. For Neyman allocation, you need to provide additional information about the strata variances. There are several ways in which you can do this, but my preferred method (because I find it the easiest way to avoid mistakes) is to specify the variances in a separate data set. This data set must:

1. Contain the variables listed in the STRATA statement of the SURVEYSELECT procedure.

2. Contain a variable named *_var_* (with the underscores!) that gives the variances (or relative variances) for each stratum. The variable must have the name *_var_* so that the procedure recognizes it as the variable containing the variances.

3. Be sorted by the variables listed in the STRATA statement.

Code 3.3 illustrates selecting a sample from *agpop* with Neyman allocation. Here, I enter conjectures for the relative variances of the strata in the data set *stratvar*: The variance in the West, for example, is set at twice the variance for the South. I then tell the SURVEYSELECT procedure that the data set *stratvar* contains the relative variance information by including the option VAR=*stratvar* in the STRATA statement.

Code 3.3. Select stratified random sample with Neyman allocation (`example0302.sas`).

```
DATA stratvar;
   INPUT region $ _var_;
   DATALINES;
S   1.0
NE  0.8
NC  1.1
W   2.0
;

PROC SORT DATA=agpop;     /* Sort the data sets. */
   BY region;
PROC SORT DATA=stratvar;
   BY region;

PROC SURVEYSELECT DATA=agpop METHOD = srs SAMPSIZE = 300 OUT = agneyman STATS SEED
     = 82835265;
   STRATA region / ALLOC = neyman VAR = stratvar;
   TITLE 'Draw stratified sample with Neyman allocation, relative variances in data
      set stratvar';
RUN;
```

Output 3.3. Select stratified random sample with Neyman allocation (`example0302.sas`).

The SURVEYSELECT Procedure

Selection Method	Simple Random Sampling
Strata Variable	region
Allocation	Neyman

Input Data Set	AGPOP
Random Number Seed	82835265
Variance Input Data Set	STRATVAR
Number of Strata	4
Total Sample Size	300
Output Data Set	AGNEYMAN

Optimal allocation. Optimal allocation requires additional information about strata variances and costs. I find it easiest to supply this information in a separate data set, which must:

1. Contain the variables listed in the STRATA statement of the SURVEYSELECT procedure.
2. Contain a variable named *_var_* (with the underscores!) that gives the variances (or relative variances) for each stratum.
3. Contain a variable named *_cost_* (with the underscores!) that gives the costs (or relative costs) for each stratum.
4. Be sorted by the variables listed in the STRATA statement.

Code 3.4 shows the code for selecting a sample from *agpop* with optimal allocation. I used the same conjectures for the relative variances of the strata as for Neyman allocation, and specified relative costs in the *_cost_* variable in the data set *stratvar2*. The STRATA statement option VAR=*stratvar2* gives the location of the data set containing the costs and variances; option COST says that the costs are also given in data set *stratvar2*. The output from the SURVEYSELECT procedure has the same form as seen for the proportional and Neyman allocations.

Code 3.4. Select stratified random sample with optimal allocation (`example0302.sas`).

```
DATA stratvar2;
   INPUT region $ _var_ _cost_;
   DATALINES;
S   1.0 1.0
NE  0.8 1.0
NC  1.1 1.4
W   2.0 1.8
;

PROC SORT DATA=agpop;     /* Sort the data sets. */
   BY region;
PROC SORT DATA=stratvar2;
   BY region;

PROC SURVEYSELECT DATA=agpop METHOD = srs SAMPSIZE = 300 OUT = agoptimal STATS SEED
     = 82835266;
   STRATA region / ALLOC = optimal VAR = stratvar2 COST;
RUN;
```

Custom allocation. You can define your own allocation by creating a data set with the desired proportion of observations for each stratum. This data set must:

1. Contain the variables listed in the STRATA statement of the SURVEYSELECT procedure.
2. Contain a variable named *_alloc_* (with the underscores!) that gives the proportion of observations to be allocated to each stratum. These proportions must sum to 1. If they do not sum to 1, you will get an error message in the log:

   ```
   ERROR: The sum of the _ALLOC_ proportions in the data set MYALLOC
   must equal 1.
   ```

3. Be sorted by the variables listed in the STRATA statement.

Code 3.5 is an example of custom allocation, for which it is desired that 40% of the observations will be taken from the West, and the remaining 60% will be divided equally among the other three strata. The total sample size of 300 is divided among the strata.

Code 3.5. Select stratified random sample with custom allocation (`example0302.sas`).

```
DATA stratalloc;
   INPUT region $ _alloc_;
   DATALINES;
S   0.2
NE  0.2
NC  0.2
W   0.4
;

PROC SORT DATA=agpop;     /* Sort the data sets. */
   BY region;
PROC SORT DATA=stratalloc;
   BY region;

PROC SURVEYSELECT DATA=agpop METHOD = srs SAMPSIZE = 300 OUT = agcustom STATS SEED
    = 82835267;
   STRATA region / ALLOC = stratalloc;
RUN;
```

You can, alternatively, specify custom sample sizes in the PROC SURVEYSELECT statement by typing SAMPSIZE=*datasetname*, where *datasetname* is a separate SAS data set containing the desired sample size in each stratum in the *_nsize_* variable (the variable must be called *_nsize_*, with the underscores). If you take this route, do not include the ALLOC option in the STRATA statement. Code 3.6 specifies that a sample with 75 observations is to be drawn from each stratum, so that the total sample size is 300.

Code 3.6. Select stratified random sample with specified sample sizes (`example0302.sas`).

```
DATA stratsize;
   INPUT region $ _nsize_;
   DATALINES;
S   75
NE  75
NC  75
W   75
;

PROC SORT DATA=agpop;     /* Sort the data sets. */
   BY region;
PROC SORT DATA=stratsize;
   BY region;

PROC SURVEYSELECT DATA=agpop METHOD = srs SAMPSIZE = stratsize OUT = agcustom STATS
     SEED = 82835267;
   STRATA region;
RUN;
```

Equal sample size in each stratum. To obtain a stratified random sample with the same sample size in each stratum, you can specify a custom allocation as in Code 3.6, in which the *_nsize_* variable of the external data set is set to the desired stratum sample size.

Alternatively, you can specify the size in each stratum in the SAMPSIZE= option. Code 3.7 will select a stratified random sample with 75 observations from each stratum (300 observations altogether). Note that no ALLOC= option is used.

Code 3.7. Select a stratified random sample with 75 observations in each stratum (`example0302.sas`).

```
PROC SURVEYSELECT DATA=agpop METHOD = srs SAMPSIZE = 75 OUT = agcustom2 STATS SEED
    = 82835267;
  STRATA region;
```

3.1.2 Additional Helpful Options for Selecting Stratified Samples

The SURVEYSELECT procedure has many other options for selecting stratified random samples. Here are two that I often find helpful in specific situations.

Minimum sample sizes for strata. By default, the SURVEYSELECT procedure selects at least 1 unit per stratum, even if the integer closest to the calculated stratum sample size is 0.

But you need at least 2 units per stratum to be able to estimate variances from a stratified sample. Placing the option ALLOCMIN=2 in the STRATA statement guarantees that each stratum will have at least 2 observations. This is shown in Code 3.8 for a proportionally allocated sample.

Code 3.8. Specifying a minimum sample size for each stratum (`example0302.sas`).

```
PROC SURVEYSELECT DATA = agpop METHOD = srs SAMPSIZE = 300 OUT = agstrat2 STATS
    SEED = 82835264;
  STRATA region / ALLOC = prop ALLOCMIN = 2;
```

You can, of course, specify the minimum allocation to be any number you want. For example, if you want to ensure that each stratum has at least 100 units, write ALLOCMIN=100. If stratum h has $N_h < 100$ and if proportional allocation is requested, that stratum will be given a sample size of $n_h = N_h$ and the remaining $(100 - N_h)$ units will be proportionally allocated to the remaining strata.

Calculate the allocation without selecting the sample. Sometimes you would like to see or tweak the allocation before selecting the sample. For example, you may want to augment the Neyman sample size in two of the strata. If you place the option NOSAMPLE in the STRATA statement, the OUT= data set will contain the sample sizes for each stratum instead of containing a selected sample.

Example 3.12 of SDA. Code 3.9 computes the proportionally allocated sample sizes displayed in Table 3.8 of SDA (and not shown here). The data set *propout* contains the sample size and the actual proportion (which may deviate from strict proportionality since sample sizes must be integers) for each stratum. The Neyman allocation can be calculated similarly, using Code 3.3 with the NOSAMPLE option.

Code 3.9. Perform the allocation without selecting the sample (example0312.sas).

```
PROC SURVEYSELECT DATA = college METHOD = srs SAMPSIZE = 200 OUT = propout SEED =
   59382;
  STRATA stratum /ALLOC = prop NOSAMPLE;
```

3.1.3 Drawing a Stratified Sample without a Population Listing

Example 3.10 of SDA. What if you do not have a data set listing the population units? Simply create your own listing in a DATA step. Code 3.10 shows you how.

Code 3.10. Create a population listing and draw stratified sample (example0310.sas).

```
DATA caribou_count;
   INPUT stratum $ count;
   DATALINES;
A 400
B  30
C  61
D  18
E  70
F 120
;

DATA caribou_pop;  /* List the population units */
   SET caribou_count;
   DO id = 1 TO count;
      OUTPUT;
   END;

DATA neyvar; /* Define the data set containing the stratum variances */
  INPUT stratum $ stdev;
  _var_ = stdev*stdev;
  DATALINES;
A 3000
B 2000
C 9000
D 2000
E 12000
F 1000
;

PROC SORT DATA=caribou_pop;  /* Sort by the stratification variable */
   BY stratum;
PROC SORT DATA=neyvar;
   BY stratum;

PROC SURVEYSELECT DATA=caribou_pop METHOD=srs SAMPSIZE=225 OUT = caribou_neyman
    SEED = 1836565293 STATS;
   STRATA stratum / ALLOC = neyman VAR = neyvar;
RUN;
```

Code 3.10 creates a data set *caribou_pop* that lists identification numbers of units in each stratum. The DO loop writes 400 records, with variable *id* taking values 1, ..., 400, for stratum A; it writes 30 records, with variable *id* taking values 1, ..., 30, for stratum B, and so on, until *caribou_pop* contains the entire population of 699 units.

The SURVEYSELECT procedure in Code 3.10 then selects a sample of size 225 using Neyman allocation, with data set *neyvar* calculating the stratum variances from the prior estimates of the within-stratum standard deviations. The output data set *caribou_neyman* contains the 225 observations that have been selected for the sample.

3.2 Computing Statistics from a Stratified Random Sample

Examples 3.2 and 3.6 of SDA. As in Chapter 2, we use the SURVEYMEANS procedure to calculate means and totals, and their standard errors and confidence intervals, from a stratified random sample. Code 3.11 shows how to find estimates for the data in `agstrat.csv`.

Code 3.11. Analysis of stratified random sample from Census of Agriculture (`example0306.sas`).

```
DATA agstrat;
   SET datalib.agstrat;
   IF acres92 < 200000 THEN lt200k = 1; /* counties with < 200000 acres in farms */
   ELSE IF acres92 >= 200000 THEN lt200k = 0;

DATA strattot;
   INPUT region $  _total_;
   DATALINES;
NE 220
NC 1054
S  1382
W  422
;

PROC SURVEYMEANS DATA = agstrat TOTAL=strattot PLOTS=ALL MEAN SUM CLM CLSUM DF;
   WEIGHT strwt;
   STRATA region;
   VAR acres92;
   TITLE 'Analysis of stratified random sample from Census of Agriculture';
RUN;
```

The syntax is similar to that in Code 2.5. What is new for stratified sampling?

TOTAL= For a stratified sample, the finite population correction (fpc) depends on the population size in each stratum. In Code 3.11, I put those four population sizes in data set *strattot*, so that the procedure knows the value of N_h for each stratum. If you use TOTAL=3078 (as we did for an SRS), the SURVEYMEANS procedure will think that each stratum has 3078 population members. If you omit the TOTAL= phrase, estimates will be computed without an fpc. If in doubt, omit TOTAL= and use the higher with-replacement standard errors.

STRATA *strat_var1 ... strat_vark* The STRATA statement gives the variable name(s) containing the stratification information (here, the stratification variable is *region*). You can include multiple variables in the STRATA statement; if you do this, the strata are formed by the cross-classification of these variables.

Output 3.11(a) shows the Data Summary and the statistics that are calculated. The first thing I look at is the sum of the weights in the Data Summary. In a stratified random sample, the sum of the weights will equal the population size if the analysis has been done correctly. Weights that sum to something else indicate that you have made a mistake either in setting up the weights or in writing the commands for the SURVEYMEANS procedure.

After that, the output is pretty self-explanatory. The estimates are those given in Examples 3.2 and 3.6 of SDA. Note that 296 degrees of freedom ($= n - H$) are used for the confidence intervals.

Output 3.11(a). Analysis of stratified random sample (`example0306.sas`).

The SURVEYMEANS Procedure

Data Summary	
Number of Strata	4
Number of Observations	300
Sum of Weights	3078

Statistics									
Variable	DF	Mean	Std Error of Mean	95% CL for Mean		Sum	Std Error of Sum	95% CL for Sum	
acres92	296	295561	16380	263325.000	327796.530	909736035	50417248	810514350	1008957721

Code 3.11 also produces a histogram and boxplot for the variable *acres92*, shown in Output 3.11(b). These estimate the histogram and boxplot for the population (see Chapter 7).

Output 3.11(b). Histogram and boxplot for *acres92* (`example0306.sas`).

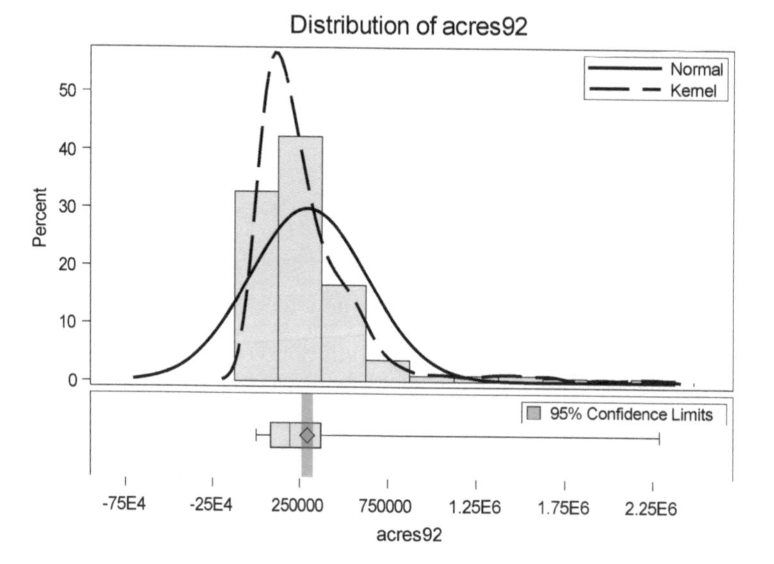

There are several ways to draw side-by-side boxplots for the strata. One way is to use the BOXPLOT procedure as in Code 3.12. The plot is similar to that in Figure 3.1 of SDA and is not shown here.

Code 3.12. Use the BOXPLOT procedure to draw side-by-side boxplots (`example0306.sas`).

```
PROC BOXPLOT DATA = agstrat;
   PLOT acres92 * region;
```

The SURVEYMEANS procedure will also draw side-by-side boxplots for the strata, along with the estimated boxplot for the population as a whole, if you include a DOMAIN statement with the stratification variable name. We shall look at the DOMAIN statement more closely in Chapter 4.

3.3 Estimating Proportions from a Stratified Random Sample

You can estimate proportions with either the SURVEYMEANS or the SURVEYFREQ procedure. Code 3.13 shows how to obtain the proportions for each category of CLASS variable *lt200k*, which was defined in the DATA statement in Code 3.11.

Code 3.13. Proportion estimates, SURVEYMEANS procedure (`example0306.sas`).

```
PROC SURVEYMEANS DATA = agstrat TOTAL=strattot PLOTS=NONE MEAN SUM CLM CLSUM;
   CLASS lt200k;
   WEIGHT strwt;
   STRATA region;
   VAR acres92 lt200k;
   TITLE 'Analysis of acres92 and lt200k from agstrat';
RUN;
```

Output 3.13. Proportion estimates, SURVEYMEANS procedure (`example0306.sas`).

Analysis of acres92 and lt200k from agstrat

The SURVEYMEANS Procedure

Data Summary	
Number of Strata	4
Number of Observations	300
Sum of Weights	3078

Class Level Information		
Variable	Levels	Values
lt200k	2	0 1

Statistics

Variable	Level	N	DF	Mean	Std Error of Mean	95% CL for Mean		Sum	Std Error of Sum	95% CL for Sum	
acres92		300	296	295561	16380	263325.000	327796.530	909736035	50417248	810514350	1008957721
lt200k	0	146	296	0.486085	0.024795	0.437	0.535	1496.170324	76.317646	1346	1646
	1	154	296	0.513915	0.024795	0.465	0.563	1581.829676	76.317646	1432	1732

Or you can use the SURVEYFREQ procedure, as shown in Code 3.14. The variable *lt200k* and the data set *strattot* were defined in Code 3.11.

Code 3.14. Proportion estimates, SURVEYFREQ procedure (`example0306.sas`).

```
PROC SURVEYFREQ DATA = agstrat TOTAL=strattot;
   WEIGHT strwt;
   STRATA region;
   TABLES lt200k/ CL CLWT;
   TITLE 'Estimating percentages from a stratified random sample';
RUN;
```

Output 3.14. Proportion estimates, SURVEYFREQ procedure (`example0306.sas`).

Estimating percentages from a stratified random sample

The SURVEYFREQ Procedure

Data Summary	
Number of Strata	4
Number of Observations	300
Sum of Weights	3078

Table of lt200k

lt200k	Frequency	Weighted Frequency	Std Err of Wgt Freq	95% Confidence Limits for Wgt Freq		Percent	Std Err of Percent	95% Confidence Limits for Percent	
0	146	1496	76.31765	1346	1646	48.6085	2.4795	43.7289	53.4881
1	154	1582	76.31765	1432	1732	51.3915	2.4795	46.5119	56.2711
Total	300	3078	5.3932E-6	3078	3078	100.0000			

The SURVEYMEANS and SURVEYFREQ procedures both calculate estimated proportions for the stratified sample. If you want to calculate confidence intervals that are based on the normal distribution, it makes no difference which procedure you use. If, however, any of the estimated proportions are close to 0 or 1, or have a skewed distribution, the SURVEYFREQ procedure allows more options for calculating confidence intervals. As with an SRS, you can calculate Clopper-Pearson confidence intervals by writing "CL (TYPE = clopperpearson)" in the TABLES statement instead of simply "CL." The procedure also calculates other asymmetric confidence intervals for proportions; see the procedure documentation for details.

3.4 Additional Code for Exercises

Some of the exercises in Chapter 3 ask you to find an ANOVA table. Here's how to do that for data set *agstrat* using the GLM procedure, which performs regression and analysis of variance. (We'll see the regression procedure that analyzes survey data in Chapter 4, and it will have a similar structure.) Note the use of the CLASS statement for the categorical variable *region*.

The MODEL statement in Code 3.15 says to fit a regression model predicting *acres92* from the categorical variable *region*, and it produces the ANOVA table shown in Output 3.15(a) for the sample. The MEANS statement requests summary statistics (means and standard deviations) for each region. Output 3.15(b), from the MEANS statement, gives the stratum summary statistics that were used in Table 3.1 of SDA. The GLM procedure also produces side-by-side boxplots for the strata.

Code 3.15. Calculating an ANOVA table with the GLM procedure (`example0306.sas`).

```
PROC GLM DATA=agstrat;
   CLASS region;
   MODEL acres92=region;
   MEANS region;
RUN;
```

Output 3.15(a). ANOVA table from MODEL statement (`example0306.sas`).

Source	DF	Sum of Squares	Mean Square	F Value	Pr > F
Model	3	7.2976366E12	2.4325455E12	27.48	<.0001
Error	296	2.6202166E13	88520829591		
Corrected Total	299	3.3499802E13			

Output 3.15(b). Summary statistics from MEANS statement (`example0306.sas`).

	acres92		
Level of region	N	Mean	Std Dev
NC	103	300504.155	172099.342
NE	21	97629.810	87449.830
S	135	211315.044	231489.714
W	41	662295.512	629433.039

3.5 Summary, Tips, and Warnings

The general structure for selecting a stratified random sample with a specified allocation using the SURVEYSELECT procedure is:

```
PROC SURVEYSELECT DATA=pop_dataset METHOD = srs SAMPSIZE = total_sample_size OUT =
    strat_sample STATS SEED = seed_integer;
  STRATA stratification_variables / ALLOC = allocation_method additional_
    allocation_info;
RUN;
```

Some types of allocation vary this syntax; see Section 3.1 for details on each type of desired allocation. Use the NOSAMPLE option in the STRATA statement if you want to calculate the allocation without selecting the sample; in that case, the OUT= data set contains the allocation instead of a selected sample.

The general structure for the SURVEYMEANS procedure with a stratified random sample is:

```
PROC SURVEYMEANS DATA=strat_sample_dataset TOTAL=stratum_totals_dataset statistics_
    keywords;
   CLASS class_var1 ... class_varp;  /* Optional */
   WEIGHT weight_variable;           /* Always include  */
   STRATA stratification_variables;  /* Always include  */
   VAR var1 var2 var3 class_var1 class_var2 class_var3;
RUN;
```

You should always include the WEIGHT, STRATA, and VAR statements when analyzing data from a stratified random sample. The procedure will produce statistics for the numeric and categorical variables listed in the VAR statement (here, three numeric and three categorical variables). The CLASS statement is optional but should be included if you are analyzing any categorical variables. Commonly used statistics keywords are listed in Table 2.1. Omit the TOTAL= statement if you want to calculate estimates without the fpc.

The general structure for the SURVEYFREQ procedure with a stratified random sample is

```
PROC SURVEYFREQ DATA=strat_sample_dataset TOTAL=stratum_totals_dataset;
   WEIGHT weight_variable;          /* Always include a weight variable */
   STRATA stratification_variables; /* Always include  */
   TABLES class_var1 class_var2 class_var3 / CL CLWT;
RUN;
```

List the variables that you want to analyze behind the TABLES statement—as many as you want. No CLASS statement is needed in the SURVEYFREQ procedure because all variables in the TABLES statement are treated as categorical.

Tips and Warnings

- Sort the population data set, and any auxiliary data sets containing stratum allocations or variances, by the stratification variable(s) before calling the SURVEYSELECT procedure. If the stratification variables are character variables, make sure the variables have the same format in all data sets. You can check the variable format using the CONTENTS procedure.
- Check the log file and the output from the SURVEYSELECT procedure to make sure that no errors were generated. Also, run the FREQ procedure to check that the sample sizes in the strata, and overall, are what you want.
- When using the SURVEYMEANS procedure with data from a stratified random sample, always include WEIGHT, STRATA, and VAR statements.
- When using the SURVEYFREQ procedure with data from a stratified random sample, always include WEIGHT, STRATA, and TABLES statements.
- In the output from the SURVEYMEANS and SURVEYFREQ procedures, check that the sum of the weights is approximately equal to the population size. If not, you have made a mistake in computing weights or in specifying the weight variable in the program.

4

Ratio and Regression Estimation

Ratio and regression estimation both use auxiliary information to increase the precision of survey estimates. This chapter shows how to incorporate that auxiliary information into survey data analyses.

4.1 Ratio Estimation

Examples 4.2 and 4.3 of SDA. The RATIO statement in the SURVEYMEANS procedure computes ratios from survey data. Let's see how it works for Examples 4.2 and 4.3 of SDA, with Code 4.1.

Code 4.1. Ratio estimation with an SRS (`example0402.sas`).

```
DATA agsrs;
   SET datalib.agsrs;
   acres92_times_xtotal = acres92*964470625;
   acres92_times_xbarU = acres92*964470625/3078;
   sampwt = 3078/300;

PROC SURVEYMEANS DATA=agsrs TOTAL=3078 MEAN STDERR CLM SUM CLSUM;
   WEIGHT sampwt;
   VAR acres92 acres87 acres92_times_xtotal acres92_times_xbarU;
   RATIO 'ratio acres92/acres87' acres92 / acres87;
   RATIO 'ratio estimator of total' acres92_times_xtotal / acres87;
   RATIO 'ratio estimator of mean' acres92_times_xbarU / acres87;
RUN;
```

You can have as many RATIO statements in a SURVEYMEANS procedure as you would like. The format of the statement is

```
RATIO 'optional label' numerator_variable / denominator_variable;
```

The label, enclosed in single quotes, is optional but helpful for locating the appropriate statistics if you are computing more than one ratio. I calculated three ratios in Code 4.1. The option CLM in the SURVEYMEANS statement requests confidence intervals for the estimated ratios.

- The first RATIO statement, requesting the ratio *acres92*/*acres87*, computes $\hat{B} = \bar{y}/\bar{x}$.
- To obtain the ratio estimator of the total, we could just compute $\hat{B}$ and then multiply by $t_x = 964{,}470{,}625$. We would also need to multiply $\text{SE}(\hat{B})$ and the confidence limits by t_x. But it is easier to just let the SURVEYMEANS procedure do the calculation by

DOI: 10.1201/9781003160366-4

defining a new variable *acres92_times_xtotal*, set equal to $y_i t_x$. Then the statement "`RATIO acres92_times_xtotal / acres87;`" produces the estimate $\hat{t}_{yr}$.

Note that the value of t_x came from the official U.S. Census of Agriculture statistics for 1987 (U.S. Bureau of the Census, 1995). This is greater than the sum of all values in data set `agpop.csv` because the data file suppresses values for small counties and thus has missing data.

- To obtain the ratio estimator $\hat{\bar{y}}_r$, I defined a new variable *acres92_times_xbarU* in the DATA step as $y_i \bar{x}_{\mathcal{U}}$. Then the statement "`RATIO acres92_times_xbarU / acres87;`" produces the estimate $\hat{\bar{y}}_r$.

Code 4.1 produces estimates of the means and totals for the variables in the VAR statement (output not shown here). Output 4.1 shows the output produced by the RATIO statements. The labels, appearing in the header of each box, help identify the output with the RATIO statement that produced it.

Output 4.1. Ratio estimation with an SRS (`example0402.sas`).

Ratio Analysis: ratio acres92/acres87					
Numerator	Denominator	Ratio	Std Error	95% CL for Ratio	
acres92	acres87	0.986565	0.005750	0.97524871	0.99788176

Ratio Analysis: ratio estimator of total					
Numerator	Denominator	Ratio	Std Error	95% CL for Ratio	
acres92_times_xtotal	acres87	951513191	5546162	940598734	962427648

Ratio Analysis: ratio estimator of mean					
Numerator	Denominator	Ratio	Std Error	95% CL for Ratio	
acres92_times_xbarU	acres87	309134	1801.871993	305587.633	312679.548

Examples 4.2 and 4.3 of SDA also explore the scatterplot of *acres92* versus *acres87*, and the correlation coefficient of the two variables. Code 4.2 calculates the correlation coefficient for *acres92* and *acres87* and draws a scatterplot. The correlation coefficient and scatterplot are given in SDA and not reprinted here. Note, however, that this code works only for a self-weighting sample. For other designs, the scatterplot must account for the survey weights; Chapter 7 will show how to do this.

Code 4.2. Correlation coefficient and scatterplot from an SRS (`example0402.sas`).

```
PROC CORR DATA = agsrs;
   VAR acres92 acres87;
   TITLE 'Correlation of acres92 and acres87';

PROC SGPLOT DATA = agsrs;
   SCATTER Y=acres92 X= acres87;
   TITLE 'Scatterplot of acres92 vs acres97';
RUN;
```

Example 4.5 of SDA. Code 4.3 shows calculating the proportion of seedlings that are still alive in 1994, for Example 4.5 of SDA. The number of trees in the population is unknown, so *sampwt* is defined to equal 1 for each observation. Output 4.3 displays the portion of the output that gives the ratio. The code in `example0405.sas` also calculates the correlation between *seed92* and *seed94* and draws the scatterplot.

Code 4.3. Ratio estimation with seedling data (`example0405.sas`).

```
PROC SURVEYMEANS DATA=santacruz PLOTS = NONE MEAN CLM STDERR VAR;
   WEIGHT sampwt;
   VAR seed94 seed92;
   RATIO 'seed94/seed92' seed94/seed92;
RUN;
```

Output 4.3. Ratio estimation with seedling data (`example0405.sas`).

Ratio Analysis: seed94/seed92						
Numerator	Denominator	Ratio	Std Error	Var	95% CL for Ratio	
seed94	seed92	0.296117	0.115262	0.013285	0.03537532	0.55685769

4.2 Regression Estimation

Example 4.7 of SDA. The SURVEYREG procedure calculates regression coefficients and regression estimators from survey data. The syntax shown in Code 4.4 is similar to that for the SURVEYMEANS procedure.

Code 4.4. Regression estimation from tree data (`example0407.sas`).

```
PROC SURVEYREG DATA=deadtrees TOTAL=100 PLOTS = ALL;
   WEIGHT treewt;
   MODEL field=photo / CLPARM DF=23;
   ESTIMATE 'Mean field trees' intercept 1 photo 11.3;
   ESTIMATE 'Total field trees' intercept 100 photo 1130;
RUN;
```

Let's look at the commands in Code 4.4, phrase by phrase:

PROC SURVEYREG requests the SURVEYREG procedure.

DATA= specifies the data set (in this case, *deadtrees*) to be analyzed, just as in the SURVEYMEANS procedure.

TOTAL= tells the size of the population (N) from which the sample was drawn. For these data, $N = 100$. As with the SURVEYMEANS procedure, if you omit the TOTAL= phrase, the estimates will be computed without the finite population correction (fpc).

PLOTS= describes which plots you want. Writing PLOTS=ALL in Code 4.4 requests a scatterplot of the data. This plot incorporates the survey weights, as will be described in Chapter 7.

WEIGHT *weight_var* says to use *weight_var* as the weight variable. My weight variable is named *treewt*, so I wrote `WEIGHT treewt`. You should always include a weight statement in the SURVEYREG procedure. If you omit it, all weights are assumed to be 1.

MODEL y = x1 x2 x3 ... xk tells which variables to include in the regression model. The dependent (y) variable precedes the "=" sign and the independent (x) variables follow it. In Code 4.4, the dependent variable is *field* and there is one independent variable, *photo*. You must include a MODEL statement in the SURVEYREG procedure, because that tells the procedure which model to fit.

CLPARM Options for the MODEL statement are given behind the slash (/). These are not required but can be helpful. CLPARM requests confidence limits for the regression parameters.

DF= Another option for the MODEL statement is to specify the degrees of freedom (df). If you omit this, the procedure uses $n-1$ df for an SRS. Since this is such a small sample, I specified $n-2$ df, which is the df associated with the estimated variance s^2. I usually omit this option for larger data sets and just let the procedure calculate the df.

ESTIMATE The ESTIMATE statement allows you to obtain estimates for predicted values from the estimated regression equation. The label (in single quotes) is optional, but helps you identify the output box corresponding to each ESTIMATE statement. For regression estimation of the mean, the statement "`ESTIMATE intercept 1 photo 11.3`" gives $\hat{B}_0(1) + \hat{B}_1(11.3)$, where $\bar{x}_{\mathcal{U}} = 11.3$. Regression estimation of the total multiplies each of these by the population size N (here, $N = 100$), and "`ESTIMATE intercept 100 photo 1130`" estimates $\hat{B}_0(100) + \hat{B}_1(1130)$, where $t_x = 1130$.

I break the output into two pieces so you can see the parts of the output that correspond to different lines in the code. Output 4.4(a) shows part the output that results from the MODEL statement. This gives the usual summary information about the number of observations and the sum of the weights, and also gives the weighted mean and sum of the dependent variable, *field*. As always, check that the sum of the weights equals the population size (which it does). The estimates of the regression coefficients are given in the last box of Output 4.4(a), with the title "Estimated Regression Coefficients." The SURVEYREG procedure also produces Fit Statistics and Tests of Model Effects (not shown here), and these will be discussed in Chapter 11.

Output 4.4(a). Data summary and regression coefficients from MODEL statement (`example0407.sas`).

Data Summary	
Number of Observations	25
Sum of Weights	100.00000
Weighted Mean of field	11.56000
Weighted Sum of field	1156.0

Estimated Regression Coefficients

Parameter	Estimate	Standard Error	t Value	Pr > |t|	95% Confidence Interval	
Intercept	5.05929204	1.42291882	3.56	0.0017	2.11576019	8.00282388
photo	0.61327434	0.12863682	4.77	<.0001	0.34716880	0.87937987

Note: The degrees of freedom for the t tests is 23.

Output 4.4(b) shows the output from the two ESTIMATE statements, each identified by its label. Note that the output gives slightly different confidence intervals for the regression estimates of the mean and total than SDA because the SURVEYREG procedure uses a slightly different (although asymptotically equivalent) formula to calculate the standard error (Example 4.7 of SDA uses formula $\hat{V}_1$ from Section 11.6 of SDA while SAS software uses formula $\hat{V}_2$ with an additional small adjustment for the number of model parameters).

Output 4.4(b). Output from ESTIMATE statements (`example0407.sas`).

Estimate								
Label	Estimate	Standard Error	DF	t Value	Pr > |t|	Alpha	Lower	Upper
Mean field trees	11.9893	0.4270	23	28.08	<.0001	0.05	11.1059	12.8726

Estimate								
Label	Estimate	Standard Error	DF	t Value	Pr > |t|	Alpha	Lower	Upper
Total field trees	1198.93	42.7014	23	28.08	<.0001	0.05	1110.59	1287.26

4.3 Domain Estimation

The SURVEYMEANS procedure allows you to calculate separate estimates for domains (population subsets for which estimates are desired) with the DOMAIN statement.

Example 4.8 of SDA. Code 4.5 shows the code for obtaining estimates in domains.

Code 4.5. Domain estimation (`example0408.sas`).

```
DATA agsrs;
   SET datalib.agsrs;
   sampwt = 3078/300;  /* sampling weight is same for each observation */
    /* Define domains of interest in variable farmcat */
   IF farms92 >= 600 THEN farmcat = "large";
   ELSE IF farms92 < 600 THEN farmcat = "small";

PROC SURVEYMEANS DATA=agsrs TOTAL = 3078 PLOTS=ALL MEAN SUM CLM CLSUM DF;
   WEIGHT sampwt;
   VAR acres92;
   DOMAIN farmcat;
RUN;
```

This is the same as the previous code to analyze data for `agsrs.csv` in Code 2.5, but with one new line:

```
   DOMAIN farmcat;
```

The DOMAIN statement requests separate estimates for each level of the variable *farmcat*, which is defined to equal "large" when *farms92* $\geq$ 600 and "small" otherwise. The variable specified in the DOMAIN statement is treated as a categorical variable, whether declared as such in a CLASS statement or not.

The VAR statement in Code 4.5 results in summary statistics for *acres92* from the entire sample (already seen in Output 2.5(b) and not shown here). The DOMAIN statement produces the statistics in Output 4.5(a).

Output 4.5(a). Domain estimation (`example0408.sas`).

The SURVEYMEANS Procedure

Statistics for farmcat Domains										
farmcat	Variable	DF	Mean	Std Error of Mean	95% CL for Mean		Sum	Std Error of Sum	95% CL for Sum	
large	acres92	299	316566	21553	274150.445	358980.857	418987302	38938277	342359512	495615092
small	acres92	299	283814	28852	227034.533	340592.894	497939808	55919525	387894116	607985499

Output 4.5(a) shows separate summary statistics for each domain specified by *farmcat*. The statistics reported are those requested in the main SURVEYMEANS statement. Note that the confidence intervals in Output 4.5(a) are slightly smaller than those given in Example 4.8 of SDA. The SURVEYMEANS procedure uses $n - 1$ df for each domain; if you have small domains in an SRS, you may want to calculate your own confidence intervals using a t distribution with $n_d - 1$ df, where n_d is the sample size in domain d.

Output 4.5(b). Side-by-side boxplots for domains (`example0408.sas`).

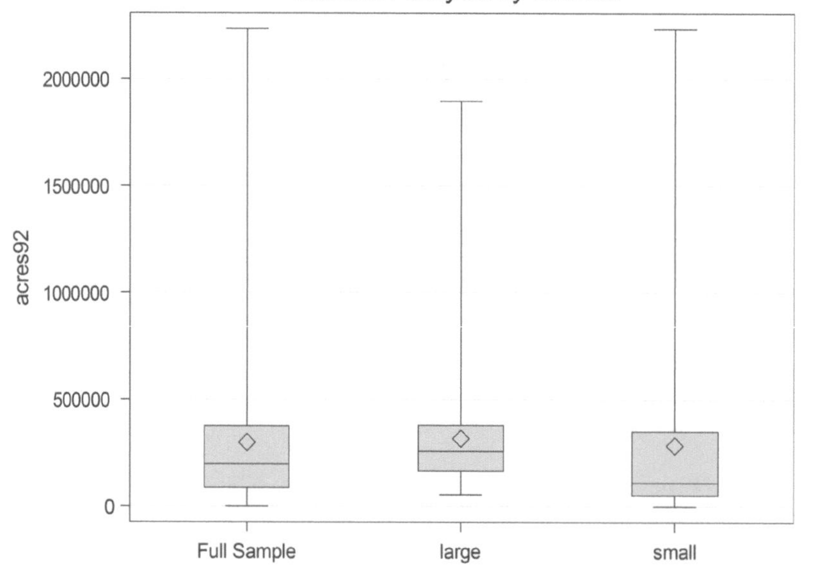

The PLOTS=ALL option results in side-by-side boxplots for the domains, shown in Output 4.5(b).

Domains in stratified samples. The DOMAIN statement also provides an easy way to obtain side-by-side boxplots for strata, if desired. These were drawn in Code 3.12 using the BOXPLOT procedure. Alternative code for drawing boxplots for the strata simply adds a DOMAIN statement to Code 3.11, as shown in Code 4.6. Here, the stratification variable is also the domain variable. Defining the strata as domains (as well as strata) tells the procedure to produce separate statistics for the four regions, and also to display side-by-side boxplots for the strata (output not shown here).

Code 4.6. Obtaining separate statistics, and side-by-side boxplots, for strata (`example0306.sas`).

```
PROC SURVEYMEANS DATA = agstrat TOTAL=strattot PLOTS=ALL MEAN SUM CLM CLSUM DF;
   WEIGHT strwt;
   STRATA region;
   VAR acres92;
   DOMAIN region;
RUN;
```

Of course, you can also analyze other domains in a stratified random sample. Simply list the variable name for which you want separate estimates in the DOMAIN statement. Code 4.7 shows how to do that to obtain separate statistics for the domain defined by variable *lt200k*, defined to equal 1 if the county has less than 200,000 acres in farms and 0 if the county has 200,000 or more acres in farms.

Code 4.7. Domain estimation in a stratified random sample (`example0306.sas`).

```
PROC SURVEYMEANS DATA = agstrat TOTAL=strattot PLOTS=ALL MEAN SUM CLM CLSUM DF;
   WEIGHT strwt;
   STRATA region;
   VAR acres92;
   DOMAIN lt200k;
RUN;
```

Why not use the BY statement? If you have used SAS software before, you may be familiar with the BY statement, which performs the calculations in the procedure separately for each level of the BY variable. For an SRS, the BY statement will give correct inferences, but for other designs, using the BY statement instead of the DOMAIN statement may produce incorrect standard errors for estimates (see Section 11.3 of SDA).

If you use a BY statement in the SURVEYMEANS procedure to calculate separate estimates, you will get a warning message:

```
NOTE: The BY statement provides completely separate analyses of the BY groups.
      It does not provide a statistically valid subpopulation or domain
      analysis, where the total number of units in the subpopulation is not
      known with certainty. If you want a domain analysis, you should include
      the DOMAIN variables in a DOMAIN statement.
```

There are circumstances in which it is appropriate to use a BY statement with the SURVEYMEANS procedure (we'll see one in Section 9.2 and discuss the issue further in Section 9.5). But in general, if you want estimates for domains, use the DOMAIN statement.

4.4 Poststratification

Example 4.9 of SDA. The SURVEYMEANS procedure computes poststratification weights and uses them to estimate population means and totals (and other statistics). The estimated population mean and total in Code 4.8 are calculated with the poststratification weights, and use variance formulas from Chapter 11 of SDA to calculate standard errors and confidence intervals.

Code 4.8. Poststratification of *agsrs* (`example0409.sas`).

```
DATA pstot;
   INPUT region $  _pstotal_;
   DATALINES;
NE 220
NC 1054
S  1382
W  422
;

PROC SURVEYMEANS DATA = agsrs TOTAL=3078 SUM CLSUM MEAN CLM;
   WEIGHT sampwt;
   VAR acres92;
   POSTSTRATA region / PSTOTAL = pstot;
   TITLE 'Poststratification of SRS from Census of Agriculture';
RUN;
```

The syntax is the same as in previous analyses of *agsrs* in the SURVEYMEANS procedure. The only feature new to Code 4.8 is the POSTSTRATA statement:

```
POSTSTRATA region / PSTOTAL = pstot;
```

The POSTSTRATA statement tells the SURVEYMEANS procedure to construct poststratification weights, using poststrata in variable *region* with poststratification totals in variable *_pstotal_* (you must use that name, with the underscores) of the data set specified by the PSTOTAL= option. Output 4.8, giving the poststratified estimates of the population mean and total of *acres92*, incorporates the poststratification in the standard errors and confidence intervals. The standard error in Output 4.8 differs slightly from that in Example 4.9 of SDA because the SURVEYMEANS procedure uses a slightly different (although asymptotically equivalent) estimator for the variance, which is discussed in Section 11.6 of SDA.

Output 4.8. Poststratification of *agsrs* (`example0409.sas`).

Poststratification of SRS from Census of Agriculture

The SURVEYMEANS Procedure

Data Summary	
Number of Poststrata	4
Number of Observations	300
Sum of Weights	3078

Statistics								
'ariable	Mean	Std Error of Mean	95% CL for Mean		Sum	Std Error of Sum	95% CL for Sum	
cres92	299778	17513	265312.879	334243.346	922717031	53906392	816633042	1028801020

You can save the poststratification weights in a data set for use in other analyses with the OUTPSWGT= option as in the following POSTSTRATA statement:

```
POSTSTRATA region / PSTOTAL = pstot OUTPSWGT = pswgt;
```

Then data set *pswgt* contains a new variable *_PSWt_* with the poststratification weight for every observation. You can use the weights in data set *pswgt* in other procedures to perform analyses, but be careful—the poststratified weights will give you point estimates, but do not contain the information needed to calculate the correct variance after poststratification. You would still need to use ratio estimation methods to calculate the variance for a poststratified estimator if using the weights in another procedure.

If, for example, you construct the poststratification weights in the SURVEYMEANS procedure, then later use those weights in the SURVEYFREQ procedure, the output from the SURVEYMEANS procedure will produce the correct standard errors that include the variance reduction from poststratification—the POSTSTRATA statement provides that procedure with the information about the poststratum membership that is needed to calculate the correct standard errors. When you use those poststratified weights in the SURVEYFREQ procedure, however, the SURVEYFREQ procedure does not know which sample members are in which poststrata and thus does not have the information that is needed to calculate the poststratified variances. Thus, the SURVEYFREQ procedure will calculate the poststratified point estimates, but the estimated variances will not account for the poststratification.

But never fear—there is a way to calculate poststratified weights in the SURVEYMEANS procedure and pass the information needed for variance estimation along to other procedures. This involves using a replication variance estimation method, as discussed in Chapter 9.

4.5 Ratio Estimation with Stratified Sampling

The RATIO statement of the SURVEYMEANS procedure always computes the ratio of variable y to variable x as $\hat{\bar{y}}/\hat{\bar{x}}$. When the STRATA statement is included, the stratification weights are used to compute the estimated means in the numerator and denominator. Thus, multiplying the ratio $\hat{\bar{y}}/\hat{\bar{x}}$ by t_x gives the combined ratio estimator.

Code 4.9 shows how to compute the ratio of *acres92* to *acres87* and the combined ratio estimator for the stratified sample in agstrat.csv. The variable *acres92_times_xtotal* is defined as *acres92**t_x as in Code 4.1. The data set *strattot*, which was created in Code 3.11, contains the population stratum sizes.

Code 4.9. Combined ratio estimation for stratified sample (ratiostrat.sas).

```
PROC SURVEYMEANS DATA = agstrat TOTAL = strattot PLOTS=ALL MEAN SUM CLM CLSUM;
   WEIGHT strwt;
   STRATA region;
   VAR acres92;
   RATIO 'Ratio of acres92 to acres87' acres92 / acres87;
   RATIO 'Ratio est of pop total' acres92_times_xtotal / acres87;
RUN;
```

Output 4.9 shows the output from the RATIO statements.

Output 4.9. Combined ratio estimation for stratified sample (`ratiostrat.sas`).

Ratio Analysis: Ratio of acres92 to acres87						
Numerator	Denominator	DF	Ratio	Std Error	95% CL for Ratio	
acres92	acres87	296	0.989997	0.006188	0.97781954	1.00217468

Ratio Analysis: Ratio est of pop total						
Numerator	Denominator	DF	Ratio	Std Error	95% CL for Ratio	
acres92_times_xtotal	acres87	296	954823130	5967910	943078218	966568041

You can calculate ratios separately for the different strata by including a `DOMAIN strat_var` statement along with the RATIO statement. This produces an overall ratio $\hat{t}_y/\hat{t}_x$ as well as separate ratios $\hat{t}_{yh}/\hat{t}_{xh}$ for each stratum h.

4.6 Model-Based Ratio and Regression Estimation

This section is optional and need only be read if covering Section 4.6 of SDA.

Example 4.11 of SDA. A model-based analysis of data from an SRS uses the same techniques taught in an introductory statistics class. Since the model-based analysis does not make use of the sampling weights, the GLM procedure, which fits general linear models for non-survey data, is used to fit the regression models and obtain the residuals. (The GLM procedure was used to calculate ANOVA tables for strata in Section 3.4.) Code 4.10 shows how to do this. The option PLOTS=FITPLOT in the GLM statement produces a scatterplot with the fitted regression line.

Code 4.10. Model-based ratio estimation (`example0411.sas`).

```
DATA agsrs;
   SET datalib.agsrs;
   IF acres87 > 0 THEN recacr87 = 1.0/acres87;
   ELSE recacr87 = .;

PROC GLM DATA=agsrs PLOTS = FITPLOT;
   MODEL acres92=acres87 / NOINT CLPARM;
   WEIGHT recacr87;
   ESTIMATE 'Pop total' acres87 964470625;
   OUTPUT OUT = resids RESIDUAL = residual PREDICTED = pred;
RUN;
```

The format for fitting a regression model in the GLM procedure is similar to that in the SURVEYREG procedure, but with one important difference: the WEIGHT statement means different things in the two procedures. In the SURVEYREG procedure, the weight variable tells how many population units are represented by each sample unit. In the GLM procedure, the weight variable contains relative weights for a weighted least squares fit.

The model used for the analysis in Code 4.10 is $Y_i = Bx_i + \varepsilon_i$, with $V(\varepsilon_i) = \sigma^2 x_i$. The model has variance proportional to x_i, so obtaining the best linear unbiased estimates under this model would use a weight value proportional to the reciprocal of the variances. This is specified by defining *recacr87 = 1 / acres87* in the DATA step. I restrict the calculation of *recacr87* to observations having *acres87* > 0 in order to avoid division by zero.

The WEIGHT statement in Code 4.10 specifies that a weighted least squares analysis is performed with weights *recacr87*. The MODEL statement has the same format as in the SURVEYREG procedure, but here the NOINT option in the MODEL statement tells that the model is to be fit without an intercept. The CLPARM option requests confidence intervals for the parameter estimates. The ESTIMATE statement requests the predicted value from the regression model when *acres87* takes on the value t_x = 964,470,625. Finally, the OUTPUT statement saves the residuals and predicted values in a data set named *resids* for further analysis.

The estimates we need for the model-based analysis are (mostly) in Output 4.10, but it takes a little bit of work to find them. First, the estimated slope, calculated using weighted least squares, is listed in the last output box, under "Parameter Estimates" for *acres87*, as 0.9865652371.

Output 4.10. Model-based ratio estimation (`example0411.sas`).

The GLM Procedure

Dependent Variable: acres92

Weight: recacr87

Source	DF	Sum of Squares	Mean Square	F Value	Pr > F
Model	1	88168461.15	88168461.15	41487.3	<.0001
Error	298	633307.00	2125.19		
Uncorrected Total	299	88801768.14			

Parameter	Estimate	Standard Error	t Value	Pr > |t|	95% Confidence Limits	
Pop total	951513191	4671509.04	203.68	<.0001	942319864.26	960706517.49

Parameter	Estimate	Standard Error	t Value	Pr > |t|	95% Confidence Limits	
acres87	0.9865652371	0.00484360	203.68	<.0001	0.9770332448	0.9960972295

The output from the ESTIMATE statement is in the box above the estimated regression slope, and is easily located from the requested label 'Pop total'. This gives the value of $\hat{t}_{yM}$ = 951,513,191. The standard error from the ESTIMATE statement, 4,671,509.04, equals $\sigma t_x / \sqrt{\sum_{i \in S} x_i}$, which is calculated without the fpc and thus is slightly higher than the standard error given in SDA.

Note that the sum of squares for error in the ANOVA table, 633,307, is the sum of squares of the weighted residuals, so the mean squared error in the ANOVA table gives $\hat{\sigma}^2 = 2125.19$.

The residuals produced by the GLM procedure are $e_i = y_i - \hat{y}_i$. For a ratio model, the weighted residuals $e_{i,wt} = e_i / \sqrt{x_i}$ should be plotted instead of e_i, because if the model variance structure is correct, the $e_{i,wt}$ should all have approximately equal variance and a plot of $e_{i,wt}$ versus the predicted values or x_i will show no patterns. Code 4.11 shows how to create the weighted residuals from the data set *resids* (which was created in the OUTPUT statement from the GLM procedure in Code 4.10) and create a plot of the weighted residuals versus the predicted values.

Code 4.11. Plotting residuals from model-based ratio estimation (`example0411.sas`).

```
DATA resids;
   SET resids;
   IF acres87 > 0  THEN  wtresid = residual/sqrt(acres87);

PROC SGPLOT DATA = resids;
    SCATTER Y = wtresid X = pred;
        XAXIS LABEL = 'Predicted value from regression model';
        YAXIS LABEL = 'Weighted residuals';
RUN;
```

The plot in Output 4.11 shows a couple of potential outliers but no other indications that the model is inappropriate.

Output 4.11. Plot of weighted residuals for Example 4.11 of SDA (`example0411.sas`).

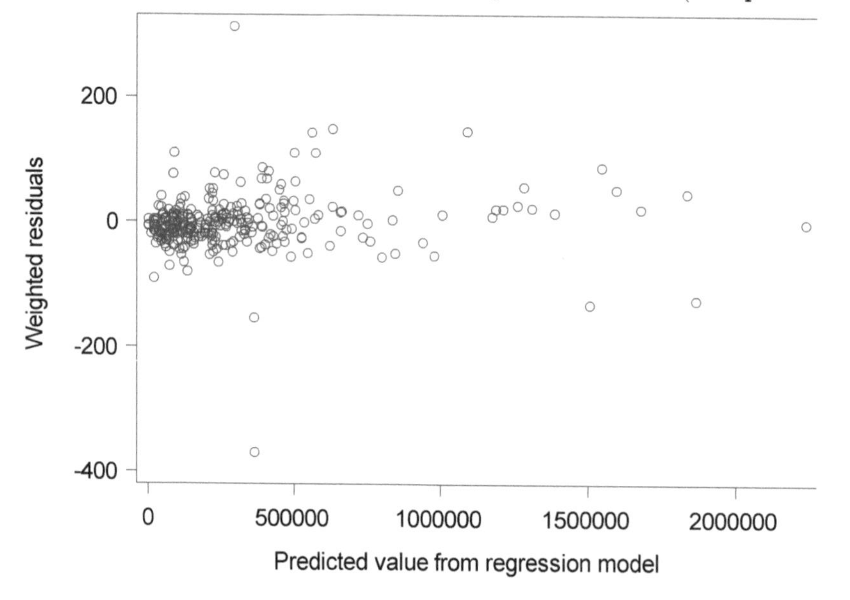

Example 4.12 of SDA. Code 4.12 gives the code and relevant output from using the GLM procedure to calculate the model-based estimates.

Code 4.12. Model-based regression estimation (`example0412.sas`).

```
PROC GLM DATA = trees PLOTS = ALL;
   MODEL field = photo/ CLPARM;
   ESTIMATE 'Pop mean' intercept 1 photo 11.3;
RUN;
```

Because the regression model is fit under the assumption that $V(\varepsilon_i) = \sigma^2$ for all observations, no WEIGHT statement is used. The ESTIMATE statement gives the standard error (without fpc) and 95% confidence interval for the regression estimator of the population mean as the predicted value $\hat{\beta}_0 + \hat{\beta}_1 \bar{x}_{\mathcal{U}}$. Output 4.12(a) displays the regression coefficients and the results from the ESTIMATE statement. The regression estimate of the population mean is 11.9893 with standard error 0.494.

Output 4.12(a). Regression coefficients and statistics from the ESTIMATE statement of the GLM procedure (`example0412.sas`).

Parameter	Estimate	Standard Error	t Value	Pr > |t|	95% Confidence Limits	
Intercept	5.059292035	1.76351187	2.87	0.0087	1.411189781	8.707394290
photo	0.613274336	0.16005493	3.83	0.0009	0.282175496	0.944373177

Parameter	Estimate	Standard Error	t Value	Pr > |t|	95% Confidence Limits	
Pop mean	11.9892920	0.49410073	24.26	<.0001	10.9671668	13.0114173

Requesting PLOTS=ALL in the GLM procedure produces a plot of y versus x as well as the full collection of residual and diagnostic plots. Output 4.12(b) displays the plot of the residuals versus *photo* (the x variable), which shows no pattern.

Output 4.12(b). Plot of residuals versus x (`example0412.sas`).

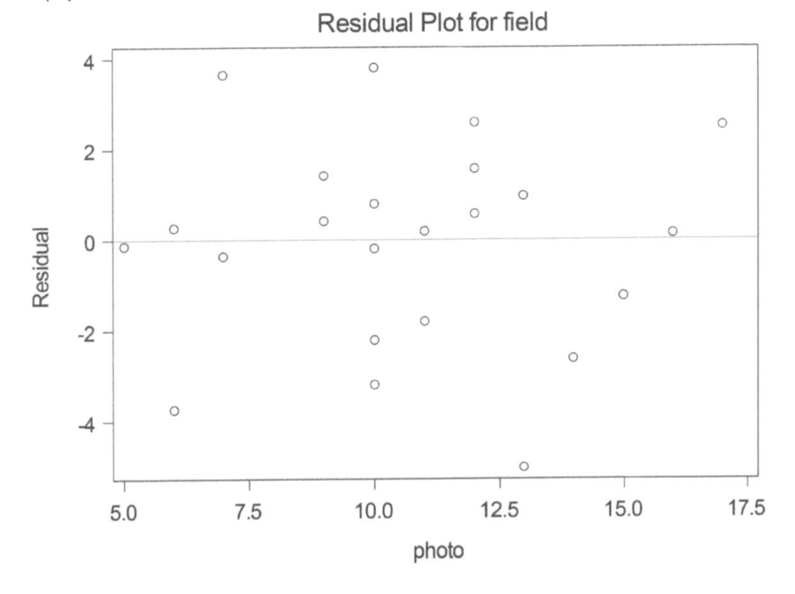

4.7 Summary, Tips, and Warnings

To calculate a ratio or ratio estimator for an SRS or stratified sample, use the RATIO statement in the SURVEYMEANS procedure. Domain estimation and poststratification are also carried out by the SURVEYMEANS procedure, through the DOMAIN and POSTSTRATA statements, respectively. The following code includes all of these features for an SRS. The CLASS, RATIO, DOMAIN, and POSTSTRATA statements are all optional.

```
PROC SURVEYMEANS DATA=srs_dataset TOTAL=pop_total statistics_keywords;
   CLASS class_var1 ... class_varp;
   WEIGHT weight_var;    /* Always include a weight variable */
   VAR var1 var2 ... vark class_var1 ... class_varp;
   RATIO  var1 / var2;   /* Ratio of the mean of var1 to mean of var2 */
   DOMAIN domain_var;    /* Domain statistics for each level of domain_var */
   POSTSTRATA poststrat_var / PSTOTAL = pstotal_data_set;
```

Regression estimators can be calculated in the SURVEYREG procedure. The general form for a straight-line regression with an SRS is:

```
PROC SURVEYREG DATA=xrs_dataset TOTAL=pop_total PLOTS = ALL;
   WEIGHT weight_variable;    /* Always include a weight variable */
   MODEL y = x / CLPARM;
   ESTIMATE 'Regression est of mean' intercept 1 x [put x̄_U here];
   ESTIMATE 'Regression est of total' intercept [put N here] x [put t_x here];
```

In the SURVEYREG procedure, the MODEL statement is required. The ESTIMATE statements are needed only if you want to obtain estimates and confidence intervals for predicted values of the regression equation, as when predicting the population mean $\bar{y}_{\mathcal{U}}$ or total t_y with regression estimation. Chapter 11 will revisit the SURVEYREG procedure for estimating population regression coefficients from survey data.

Tips and Warnings

- Draw a scatterplot of your data when fitting a ratio or regression model, so you can see whether ratio or regression estimation is likely to improve efficiency.
- Always include a WEIGHT statement in the SURVEYMEANS and SURVEYREG procedures. If the sample is stratified, also include a STRATA statement.
- Ratios, and ratio estimates of population means and totals, can be calculated with the RATIO statement in the SURVEYMEANS procedure. If you are estimating more than one ratio, include a label in each RATIO statement so you will be able to identify them in the output.
- Always include a MODEL statement in the SURVEYREG procedure. This tells the procedure which variables should be in the model. The ESTIMATE statement gives predicted values for the estimated regression function; include a label in each ESTIMATE statement so you can easily identify it in the output.
- For domain estimation, use the DOMAIN statement in the SURVEYMEANS or SURVEYREG procedure. It may be tempting to calculate statistics for a subset of the sample by either (1) creating a data set containing only that subset or (2) using a BY statement, but those options can result in incorrect standard errors.
- Poststratification weights can be calculated in the SURVEYMEANS procedure through the POSTSTRATA statement.

5

Cluster Sampling with Equal Probabilities

This chapter presents the commands for computing estimates from one-stage and multi-stage cluster samples. These commands are the same whether clusters are selected with equal probabilities (Chapter 5 of SDA) or with unequal probabilities (Chapter 6 of SDA).

Chapter 6 will tell how to select a one-stage or two-stage cluster sample with equal or unequal probabilities. The syntax is similar for both, and deferring the sample selection examples to Chapter 6 allows us to look at the general case.

5.1 Estimating Means and Totals from a Cluster Sample

5.1.1 One-Stage Cluster Sampling

Example 5.2 of SDA. Code 5.1 finds estimates for the population mean and total in the GPA data.

Code 5.1. Calculate estimates for the GPA example (`example0502.sas`).

```
PROC SURVEYMEANS DATA = gpa TOTAL = 100 NOBS MEAN SUM CLM CLSUM;
   WEIGHT wt;
   CLUSTER suite;
   VAR gpa;
RUN;
```

As always, the WEIGHT statement gives the weight variable to be used when analyzing the data. This is the weight at the observation-unit level. Here, variable *wt* has value 20 for each student. The CLM and CLSUM options request confidence intervals for the estimates.

The CLUSTER statement tells the SURVEYMEANS procedure that variable *suite* is the clustering variable. If you omit the CLUSTER statement, the data will be (incorrectly) analyzed as an SRS. Note that in a cluster sample, TOTAL= refers to the total number of primary sampling units (psus) in the population (100), not the total number of secondary sampling units (ssus) in the population, which is 400. Including the TOTAL= option tells the procedure to include the finite population correction (fpc) when calculating variances. This produces the variance estimates of estimated population means and totals given in Section 5.2 of SDA.

The Data Summary in Output 5.1 shows that the data have 5 psus and 20 observations, with the weights summing to the number of students in the population, 400. This is the only indication in the output that the clustering was used in the analysis. Otherwise, the form of the Statistics output is the same as for simple random or stratified sampling. Because

DOI: 10.1201/9781003160366-5

this is a cluster sample, confidence intervals are calculated using a t percentile with degrees of freedom (df) equal to the number of psus minus 1.

Output 5.1. Calculate estimates for the GPA example (`example0502.sas`).

The SURVEYMEANS Procedure

Data Summary	
Number of Clusters	5
Number of Observations	20
Sum of Weights	400

Statistics									
Variable	N	Mean	Std Error of Mean	95% CL for Mean		Sum	Std Error of Sum	95% CL for Sum	
gpa	20	2.826000	0.163665	2.37159339	3.28040661	1130.400000	65.465961	948.637354	1312.16265

When using the SURVEYMEANS procedure with cluster samples, always check that the number of clusters in the output equals the number of psus in your sample, and that the sum of the weights is a reasonable value for the population size. In general, when clusters have unequal sizes, the sum of the weights for the sample is an estimate of the population size; if you took a different sample, you would get a different sum for the weights. But you can check whether the sum of the weights is close to the population size. For example, for a survey of the U.S. population (which, in 2020, was approximately 330 million people), the sum of the sampling weights for your sample might be 326 million—not exactly the population value, but in the right neighborhood. If the sum of your sampling weights is 3500, then it is likely that you made a mistake somewhere when computing or using the weights.

The basic syntax in Code 5.1 is used for all cluster samples. It does not matter whether the psus are selected with equal or unequal probabilities, or how many stages of sampling are done. The SURVEYMEANS procedure uses only the first-stage clustering information to calculate variances.

Example 5.6 of SDA. The syntax for analyzing the data from a cluster sample with unequal sizes is exactly the same, as shown in Code 5.2. Again, note that 11 df (number of psus minus 1) are used. Because a one-stage sample is taken, the sampling weight is $N/n = 187/12$ for each observation.

Code 5.2. Calculate estimates for Example 5.6 of SDA (`example0506.sas`).

```
DATA algebra;
   SET datalib.algebra;
   sampwt = 187/12;

PROC SURVEYMEANS DATA = algebra TOTAL = 187 NOBS MEAN DF CLM SUM CLSUM;
   WEIGHT sampwt;
   CLUSTER class;
   VAR score;
RUN;
```

Output 5.2. Calculate estimates for Example 5.6 of SDA (`example0506.sas`).

The SURVEYMEANS Procedure

Data Summary	
Number of Clusters	12
Number of Observations	299
Sum of Weights	4659.41667

Statistics										
Variable	N	DF	Mean	Std Error of Mean	95% CL for Mean		Sum	Std Error of Sum	95% CL for Sum	
score	299	11	62.568562	1.491578	59.2856211	65.8515026	291533	19893	247749.374	335316.626

5.1.2 Multi-Stage Cluster Sampling

SAS software can be used to calculate without-replacement variances for two-stage cluster samples in which an SRS is selected at each stage, but it is easier in the SURVEYMEANS procedure to calculate the with-replacement variance. Let's first discuss the with-replacement variance estimation for Example 5.8 of SDA and then show some options for without-replacement sampling using Example 5.7.

Example 5.8 of SDA. The syntax for two-stage cluster sampling in the SURVEYMEANS procedure is exactly the same as that for one-stage cluster sampling, as you can see in Code 5.3. The WEIGHT variable contains the sampling weights (for this example, the relative weights *relwt* are used), and the CLUSTER statement gives the name of the variable that specifies the psu membership.

Code 5.3. Calculate estimates for coots data (`example0508.sas`).

```
DATA coots;
   SET datalib.coots;
   relwt = csize/2;

PROC SURVEYMEANS DATA = coots DF MEAN CLM CV;
   WEIGHT relwt;
   CLUSTER clutch;
   VAR volume;
   TITLE 'Calculate mean egg volume, CI, and CV';
RUN;
```

Output 5.3. Calculate estimates for coots data (`example0508.sas`).

Data Summary	
Number of Clusters	184
Number of Observations	368
Sum of Weights	1758

Statistics						
Variable	DF	Mean	Std Error of Mean	95% CL for Mean		Coeff of Variation
volume	183	2.490778	0.061001	2.37042271	2.61113374	0.024491

Output 5.3 shows the estimates given in Example 5.8 of SDA. Note that 183 df (number of psus minus one) are used for the confidence interval.

The PROC SURVEYMEANS statement in Code 5.3 does not contain the TOTAL= option. This is because the total number of clutches in the population, N, is unknown. Consequently, the procedure does not use an fpc when calculating estimates. In general, I recommend omitting the TOTAL= option for multi-stage cluster sampling even when N is known, however, and the remainder of this section discusses why.

Finite population corrections in two-stage cluster sampling. The SURVEYMEANS procedure calculates the estimated totals and means for a cluster sample using the sampling weights. But the weights do not contain sufficient information for calculating the variance—that requires knowledge of the sampling design.

The without-replacement variance estimates for two-stage cluster sampling given in Section 5.3 of SDA require knowledge not only of the psu membership of each observation, but also of the population size M_i and sample size m_i for each psu in the sample. If a survey has three stages, the without-replacement variance estimate requires knowledge of the psu membership, population size, and sample size; the ssu membership, population size, and sample size; and the tertiary sampling unit membership, population size, and sample size. It can be extremely complicated to keep track of all this information.

The SURVEYMEANS procedure, and other SAS software procedures for survey data analysis, use only the information at the first stage of sampling. You supply the psu-level clustering information in the CLUSTER statement, and you do not provide information on ssu membership, or numbers of ssus in the population and sample, to the procedure. Although the procedure does not compute the exact without-replacement variance for two-stage cluster sampling given in Section 5.3 of SDA, it computes a close approximation with the benefit that you do not have to keep track of information at the ssu and subsequent levels. Chapter 6 outlines additional benefits of the approach taken in the SURVEYMEANS procedure when the psus are selected with unequal probabilities.

Here are two options for estimating the variance of estimated means and totals in without-replacement two-stage sampling.

Option 1. Estimate the with-replacement variance (recommended). As shown in Sections 5.3 and 6.6 of SDA, the estimated variability among estimated psu totals, s_t^2, also includes variability from the subsequent stages of sampling. If you estimate the with-replacement variance (at the psu level), the variance estimator incorporates *all* the variability from subsequent stages of sampling. The expected value of the with-replacement variance estimator is larger than the true variance of the without-replacement sample, but the difference is small if the sampling fraction at the psu level, n/N, is small.

To estimate the with-replacement variance, simply call the procedure without including TOTAL= in the PROC SURVEYMEANS statement. Do the same for other survey analysis procedures such as the SURVEYFREQ or SURVEYREG procedures.

Option 2. Do additional calculations. If you include TOTAL= in the SURVEYMEANS statement, the variance estimate provided by the procedure incorporates the fpc at the psu level but it does not include the variability at the ssu level. The SURVEYMEANS procedure with TOTAL= gives the first term in the estimator

$$\hat{V}\left(\hat{t}_{\text{unb}}\right) = N^2\left(1-\frac{n}{N}\right)\frac{s_t^2}{n} + \frac{N}{n}\sum_{i\in\mathcal{S}}\left(1-\frac{m_i}{M_i}\right)M_i^2\frac{s_i^2}{m_i}. \tag{5.1}$$

If you have the information on the ssu-level population sizes M_i, sample sizes m_i, and sample variances s_i^2, you can then calculate the second term of the variance estimator separately and add it to the variance produced by the SURVEYMEANS procedure. This is usually not worth the effort, but can result in a smaller variance than Option 1 when the number of psus in the population, N, is small.

Note that if you do not go through the extra step of computing the ssu-level term in the variance estimate, but you include TOTAL= in the SURVEYMEANS statement, the variance estimate is too small because it does not include the second term in (5.1). Of course, if n/N is small, the second term in (5.1) is small relative to the first term, but in that situation, the difference between the without-replacement variance in (5.1) and the with-replacement variance in Option 1 is also small. With more than two stages of clustering, you need to include extra terms for the variance at all stages of sampling to calculate the without-replacement variance.

In general, I recommend calculating the with-replacement variance (omitting TOTAL=) for multi-stage cluster sampling. It produces a variance estimate whose expectation is slightly larger than the true variance, but if n/N is small, the difference is negligible. If forced to choose between a standard error that is slightly too large and one that is too small, I usually prefer the former because a too-small standard error leads to claiming that estimates are more precise than they really are.

The most important thing to keep in mind for computing standard errors for cluster samples is that ssus in the same psu are usually more homogeneous than randomly selected ssus from the population. Thus, the essential feature for calculating standard errors is to capture that homogeneity by including the CLUSTER statement in the SURVEYMEANS procedure. The issue of "to fpc or not to fpc" is minor compared with the effect of clustering.

Example 5.7 of SDA. If you want to do the extra work, you can use SAS software to calculate the without-replacement variance when an SRS is taken at both stages. Let's illustrate with the schools data from Example 5.7 of SDA.

First, let's do the with-replacement variance calculations, in Code and Output 5.4.

Code 5.4. With-replacement variance estimation for schools (`example0507.sas`).

```
PROC SURVEYMEANS DATA = schools MEAN SUM DF CLM CLSUM;
   WEIGHT finalwt;
   CLUSTER schoolid;
   VAR math;
   TITLE 'With-replacement variance calculations';
RUN;
```

Output 5.4. With-replacement variance estimation for schools (`example0507.sas`).

Statistics										
Variable	DF	Mean	Std Error of Mean	95% CL for Mean		Sum	Std Error of Sum	95% CL for Sum	
math	9	33.122948	1.759894	29.1417926	37.1041040	572116	55524	446512.948	697719.302

The calculations in Code 5.5 to find the first term of the without-replacement variance in (5.1) are similar, except we add the TOTAL= option to capture the $(1 - n/N)$ factor. Note that the standard error in Output 5.5 is $\sqrt{1 - 10/75}$ multiplied by the with-replacement standard error in Output 5.4.

Code 5.5. Without-replacement variance estimation for schools: term 1 (`example0507.sas`).

```
PROC SURVEYMEANS DATA = schools TOTAL=75 MEAN SUM DF CLM CLSUM;
   WEIGHT finalwt;
   CLUSTER schoolid;
   VAR math;
   TITLE 'Term 1 of without-replacement variance calculations';
RUN;
```

Output 5.5. Without-replacement variance estimation for schools: term 1 (`example0507.sas`).

Statistics									
Variable	DF	Mean	Std Error of Mean	95% CL for Mean		Sum	Std Error of Sum	95% CL for Sum	
math	9	33.122948	1.638372	29.4166941	36.8292026	572116	51690	455185.931	689046.319

But the standard error in Output 5.5 does not include the second term of (5.1). We can calculate that by finding the summary statistics for each psu, as in Table 5.7 of SDA, with Code 5.6.

Code 5.6. Without-replacement variance estimation for schools: term 2 (`example0507.sas`).

```
PROC SORT DATA = schools;
   BY schoolid;
PROC MEANS DATA=schools MEAN VAR NOPRINT;
   BY schoolid;
   VAR Mi math;
   OUTPUT OUT = schoolout MEAN = Mi ybari VAR(math) = s2i;
   TITLE 'Calculate summary statistics for psus';

DATA schoolout;
   SET schoolout (KEEP = schoolid Mi ybari s2i);
   thati = ybari*Mi;
   term2vari = (1-20/Mi)*(Mi**2)*s2i/20;
   TITLE 'Calculate the quantities needed for variance formulas';

PROC MEANS DATA=schoolout MEAN SUM;
   VAR Mi thati term2vari;
   TITLE 'Calculate means and sums for psu summary statistics';
RUN;
```

Output 5.6. Without-replacement variance estimation for schools: term 2 (`example0507.sas`).

Calculate means and sums for psu summary statistics

The MEANS Procedure

Variable	Mean	Sum
Mi	230.3000000	2303.00
thati	7628.22	76282.15
term2vari	290208.74	2902087.43

Output 5.6 gives the sums of the columns of Table 5.7 of SDA, and we can now plug into the formulas to get the without-replacement standard error as

$$\sqrt{1.638372^2 + \frac{1}{(10)(75)(230.3)^2}2902087} = \sqrt{2.684 + 0.073} = \sqrt{2.757} = 1.66.$$

5.2 Estimating Proportions from a Cluster Sample

As always, you can estimate proportions with either the SURVEYMEANS procedure (by including the variable for which proportions are desired in the CLASS statement) or the SURVEYFREQ procedure. It should be no surprise at this point that the SURVEYFREQ procedure for cluster samples has the same syntax as the SURVEYMEANS procedure. You simply add a CLUSTER statement that specifies the variable containing the psus.

Example 5.7 of SDA. Let's look at the SURVEYFREQ procedure for the schools data, to estimate the percentage and total number of students with *mathlevel*=2.

Code 5.7. Estimate percentage of students in each *mathlevel* category (`example0507.sas`).

```
PROC SURVEYFREQ DATA = schools;
   WEIGHT finalwt;
   CLUSTER schoolid;
   TABLES mathlevel / CL CLWT;
   TITLE 'With-replacement variance calculations for mathlevel';
RUN;
```

Output 5.7. Estimate percentage of students in each *mathlevel* category (`example0507.sas`).

Table of mathlevel

mathlevel	Frequency	Weighted Frequency	Std Err of Wgt Freq	95% Confidence Limits for Wgt Freq		Percent	Std Err of Percent	95% Confidence Limits for Percent	
1	132	12303	2244	7227	17380	71.2310	5.4168	58.9774	83.4846
2	68	4969	676.25819	3439	6499	28.7690	5.4168	16.5154	41.0226
Total	200	17273	2095	12534	22011	100.0000			

The confidence interval produced by Code 5.7 is of the form

percentage ± (t critical value) (standard error of percentage) where the with-replacement standard error is calculated the same way as in the SURVEYMEANS procedure, assuming that the estimated population proportion $\hat{p}$ is approximately normally distributed. When the population proportion is close to zero or one, however, the distribution of $\hat{p}$ is skewed. When p is close to zero or one, a nominal 95% confidence interval based on the normal distribution may not be accurate, in the sense that if you were able to repeatedly take samples from the population and calculate a confidence interval from each one, you would find that fewer than 95% of the intervals included the true population proportion. In other words, when p is close to zero or one, the actual coverage probability of a 95% normal-distribution-based confidence interval will be less than 95%.

The SURVEYFREQ procedure will calculate alternative confidence intervals that have more accurate coverage probability when the population proportion is close to zero or one. These include the Clopper-Pearson interval discussed in Section 2.4 and the Wilson confidence interval. Simply add the CL(TYPE=) option in the TABLES statement.

5.3 Model-Based Design and Analysis for Cluster Samples

We often use models when designing a cluster sample, as shown in Section 5.4 of SDA. Data from a previous survey or pilot sample may be used to estimate the optimal subsampling or psu size. This often involves estimating the value of R^2 or R_a^2, which can be obtained from an ANOVA table.

Example 5.12 of SDA. Code 5.8 shows how to calculate an ANOVA table for the schools data. The MEANS statement is optional; it displays summary statistics for each school.

Code 5.8. Calculate ANOVA table for schools data (`example0507.sas`).

```
PROC GLM DATA = schools;
   CLASS schoolid;
   MODEL math = schoolid;
   MEANS schoolid;
RUN;
```

Output 5.8. Calculate ANOVA table for schools data (`example0507.sas`).

Source	DF	Sum of Squares	Mean Square	F Value	Pr > F
Model	9	7018.48000	779.83111	7.58	<.0001
Error	190	19538.40000	102.83368		
Corrected Total	199	26556.88000			

R-Square	Coeff Var	Root MSE	math Mean
0.264281	29.25763	10.14069	34.66000

Example 5.14 of SDA. A model-based analysis employs a random effects model, in which the psu means are assumed to be normally distributed random variables with mean μ.

In SAS/STAT software, the MIXED procedure calculates estimates from random effects models and general mixed models. The syntax is similar to that from the GLM procedure in Code 5.8, except that in Code 5.9 for mixed models, the random effects are specified in the RANDOM statement and omitted from the MODEL statement. Thus, the MODEL statement has no x variables for this example since the only fixed effect is the mean, and *schoolid*, the variable describing the psu membership, is placed in the RANDOM statement. The estimate for the mean and an associated 95% confidence interval are requested with the SOLUTION and CL options in the MODEL statement.

Code 5.9. Fit random effects model to schools data (`example0514.sas`).

```
PROC MIXED DATA = schools;
   CLASS schoolid;
   MODEL math = / SOLUTION CL;
   RANDOM schoolid;
RUN;
```

Output 5.9(a) shows the part of the output from the MIXED statement that gives the variance component estimates $\hat{\sigma}_A^2 = 33.85$ and $\hat{\sigma}^2 = 102.83$, and Output 5.9(b) gives the estimated mean and its standard error and confidence interval. You can see the rest of the output from the MIXED procedure (not needed for this example) by running the full code in `example0514.sas`.

Output 5.9(a). Estimated variance components from model-based analysis of schools data (`example0514.sas`).

Covariance Parameter Estimates	
Cov Parm	Estimate
schoolid	33.8499
Residual	102.83

Output 5.9(b). Estimated mean from model-based analysis of schools data (`example0514.sas`).

Solution for Fixed Effects								
Effect	Estimate	Standard Error	DF	t Value	Pr > |t|	Alpha	Lower	Upper
Intercept	34.6600	1.9746	9	17.55	<.0001	0.05	30.1931	39.1269

A model-based analysis predicts the values of observations that are not observed in the data. For this data set, the unobserved values are the students that are not measured in the sampled schools, as well as all of the unsampled schools in the population. The estimated mean from the MIXED procedure does not account for the population sizes of the different schools, and gives a different value than in Example 5.7 of SDA. An alternative model-based estimator would account for the unequal sizes of the clusters and predict the values of the unsampled students in each sampled school.

The MIXED procedure will also perform model diagnostics if you request the INFLUENCE and RESIDUAL options in the MODEL statement.

5.4 Additional Code for Exercises

Exercise 5.40 of SDA. The exercise uses the macro *intervals*, available on the book website in file `intervals.sas`. A macro is a collection of SAS statements that can be executed by typing the macro name, giving you the ability to execute the same set of statements on multiple data sets and variables. In this exercise, you will execute a macro that has already been written; Slaughter and Delwiche (2004) give a short introduction to macro programming for readers who want to learn how to write their own macros.

To run the *intervals* macro, first put an %INCLUDE statement in your program that gives the filepath on your computer to the program `intervals.sas`, as shown in Code 5.10. When you run the %INCLUDE statement, SAS executes the code in the file `intervals.sas`, which in this case loads the macro *intervals*. You then run the macro by typing `%intervals` with the desired arguments.

Code 5.10. Load and run the *intervals* macro (`run_intervals.sas`).

```
%INCLUDE "C:\MyFilePath\intervals.sas"; /* Load the macro */

%intervals(icc = .3); /* Run the macro with ICC = .3 */
```

After loading the macro *intervals*, you can run it as many times as you want. The arguments of the macro are given in Table 5.1.

TABLE 5.1
Arguments of the *intervals* macro.

Argument	Description
ICC	Desired intraclass correlation coefficient, must be between 0 and 1 (default is 0).
OUTDATA	Name of output data set containing results from individual interval estimates (default is *cistats*).
NUMINTERVALS	Number of confidence intervals to be generated (default is 100).
CLUSTERSIZE	Number of observations, M, in each population cluster (default is 5).
SAMPSIZE	Number of clusters to be sampled (default is 10).
POPSIZE	Number of clusters in population (default is 5000). This should be set to be at least 200 times as large as SAMPSIZE so that the fpc is negligible.
MU	Population mean (default is 0).
SIGMA	Population standard deviation (default is 1).
PLOTS	Set to yes if it is desired to draw graphs of the confidence intervals (default is yes).

For the exercise, you are asked to generate 100 intervals of 50 observations each, taken in 10 clusters of size 5. This uses the default values of all arguments except for ICC. When running the macro, you need to specify only the arguments that differ from the default values, so that you can generate 100 intervals with ICC = 0.3 by running either of the statements in the following code. The two macro calls are equivalent; the second call explicitly requests the default values of arguments.

```
/* The following two macro calls are equivalent */
%intervals(icc = .3);
%intervals(icc = .3, outdata= cistats, numintervals= 100, clustersize= 5, sampsize=
     10, popsize= 5000, mu= 0, sigma= 1, plots= yes);
```

The macro calculates two sets of interval estimates: a set that uses SRS formulas and hence has coverage probability less than 0.95 when ICC $\neq 0$, and a second set that calculates the correct confidence intervals using the formulas for one-stage cluster sampling. It prints the proportion of intervals that include the true population mean, and the average width of the interval estimates, for the two methods. Finally, if PLOTS=yes, it produces two graphs. The first graph shows the interval estimates produced for each sample if analyzed as an SRS, and the second shows the interval estimates produced for each sample when analyzed as a cluster sample.

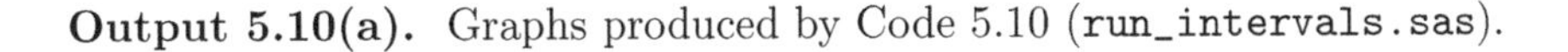

Output 5.10(a). Graphs produced by Code 5.10 (`run_intervals.sas`).

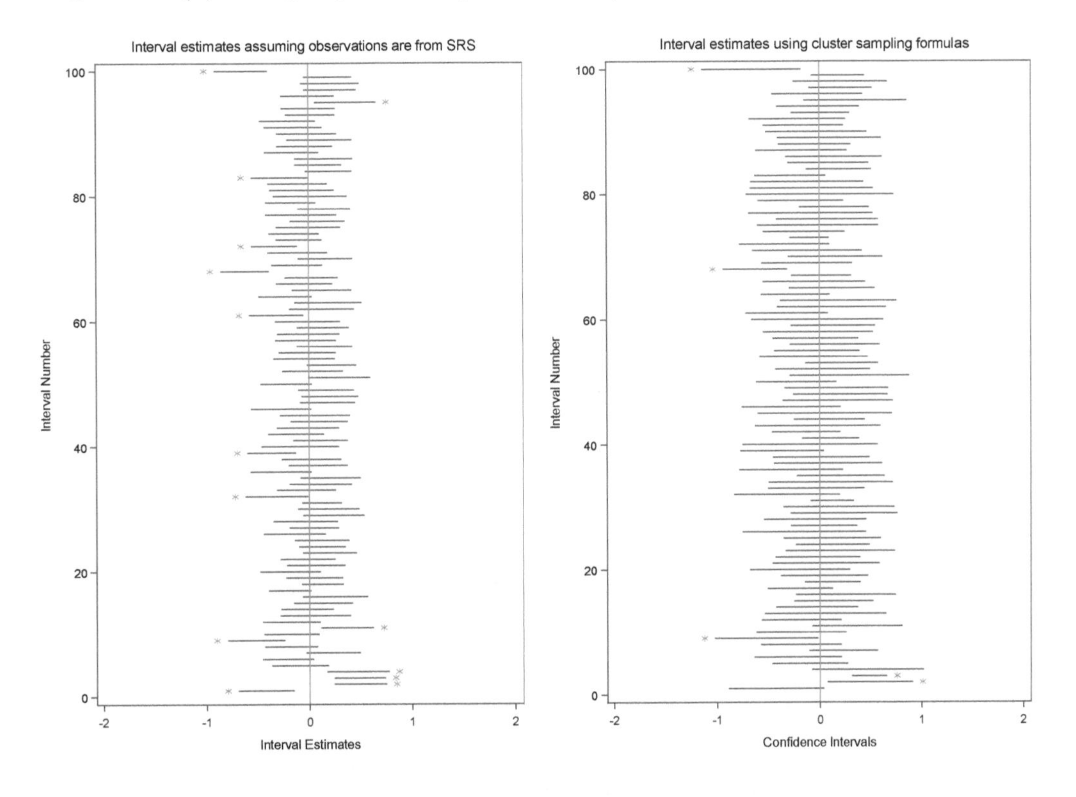

Output 5.10(a) shows the two graphs for this example. The interval estimates that do not include the population mean are marked by asterisks. The estimated coverage probability for each procedure is thus the proportion of intervals that do not have asterisks. For the simulation shown in Output 5.10(a), using the (incorrect) SRS formulas to calculate the confidence intervals for this resulted in an estimated coverage probability of 0.86, compared with the estimated coverage probability of 0.95 when the cluster formulas are used.

Output 5.10(b) shows the summary statistics for this simulation. Note that the macro uses a random number seed generated by the internal clock; you will get different intervals each time you run the program.

Output 5.10(b). Summary statistics from Code 5.10 (`run_intervals.sas`).

Estimated Coverage Probabilities and Mean CI Widths

Number_of_intervals	SRS_cov_prob	Cluster_cov_prob	SRS_mean_CI_width	Cluster_mean_CI_width
100	0.86	0.95	0.55207	0.89971

The data set specified by the OUTDATA argument to the macro (for this example, *cistats*) contains additional information on the individual interval estimates generated by the macro. The same sample is used to calculate the SRS-based and cluster-based interval estimates for each Replicate k ($k = 1, \ldots, 100$), so if you want to test whether the widths for the two types of intervals are significantly different, you need to use a paired t test. Output 5.10(c) shows the first five observations from data set *cistats*, with the mean, lower (lcl) and upper (ucl) confidence limits, and width for each interval, as well as an indicator variable that equals 1 if the interval contains the true value of the mean and 0 otherwise.

Output 5.10(c). First five observations of *cistats* (`run_intervals.sas`).

Summary Statistics from Individual Interval Estimates

Obs	Replicate	srs_mean	srs_lcl	srs_ucl	in_srs_ci	srs_ciwidth	clus_mean	clus_lcl	clus_SE	clus_ucl	in_clus_ci	clus_ciwidth
1	1	-0.42528	-0.69864	-0.15192	0	0.54672	-0.425276	-0.8906064	0.205702	0.0400535	1	0.93066
2	2	0.49241	0.23481	0.75000	0	0.51519	0.492406	0.0746899	0.184654	0.9101230	0	0.83543
3	3	0.48771	0.23655	0.73888	0	0.50232	0.487715	0.3179264	0.075056	0.6575035	0	0.33958
4	4	0.47011	0.16686	0.77335	0	0.60649	0.470106	-0.0792671	0.242854	1.0194786	1	1.09875
5	5	-0.09402	-0.37127	0.18323	1	0.55450	-0.094024	-0.4668201	0.164797	0.2787729	1	0.74559

5.5 Summary, Tips, and Warnings

Clusters in the SURVEYMEANS and SURVEYFREQ procedures are specified in the CLUSTER statement. The general form of the SURVEYMEANS procedure for a cluster sample (without stratification) is:

```
PROC SURVEYMEANS DATA=cluster_sample_dataset TOTAL=cluster_total statistics_
   keywords;
  CLASS class_var1 ... class_varp;  /* Optional; include the CLASS statement if
   analyzing any categorical variables */
  WEIGHT weight_variable;     /* Always include a weight variable */
  CLUSTER cluster_variable;  /* Always include  */
  VAR var1 var2 ... vark class_var1 ... class_varp;
  /* var1 var2 ... vark are the numeric variables to be analyzed */
  /* class_var1 ... class_varp are the categorical variables to be analyzed */
```

The general form of the SURVEYFREQ procedure for a cluster sample (without stratification) is:

```
PROC SURVEYFREQ DATA=cluster_sample_dataset TOTAL=cluster_total;
  WEIGHT weight_variable;     /* Always include a weight variable */
  CLUSTER cluster_variable;  /* Always include  */
  TABLES class_var1 ... class_varp / CL CLWT;
  /* class_var1 ... class_varp are the categorical variables to be analyzed */
```

The TOTAL= option calculates statistics with a psu-level fpc, $(1 - n/N)$. If you use this option, specify TOTAL=N, the number of psus in the population; do not use the total number of observation units. In two-stage equal-probability sampling, the SURVEYMEANS and SURVEYFREQ procedures with the TOTAL= option calculate standard errors that are slightly too small because they do not include all of the variability at the second stage of sampling; it is usually preferable to calculate the with-replacement variance (omitting the TOTAL= option) instead.

Tips and Warnings

- Always include the WEIGHT statement in your analysis. The WEIGHT variable contains the weight for each record (observation unit) in the data set.
- In the output from the SURVEYMEANS and SURVEYFREQ procedures, check that the sum of the weights is approximately equal to the population size.
- If there is clustering in the survey, include the CLUSTER statement. This guarantees that the SURVEYMEANS, SURVEYFREQ, and other survey analysis procedures take the clustering into account when calculating standard errors. Check that the number of clusters listed in the Data Summary equals the number of psus in your sample.
- For cluster samples, TOTAL= refers to the number of psus in the population, not to the number of observation units. I usually omit TOTAL= when analyzing data from multi-stage cluster samples and calculate variances without the fpc.
- The most important consideration when calculating confidence intervals for estimates from a cluster sample is to include the clustering. If you do not account for the clustering, confidence intervals will usually be much smaller than they should be. By contrast, the other decisions—whether to use the with-replacement or without-replacement estimate of the variance, or choice of degrees of freedom—usually have much smaller effect on the size of the confidence intervals.

6

Sampling with Unequal Probabilities

Chapter 5 showed how to calculate estimated means and totals from samples in which clusters are sampled with equal probabilities. The syntax for computing estimates from samples selected with unequal probabilities is the same; the weights and clustering give the survey analysis procedures the information needed to calculate the estimates. Section 6.3 shows the code used to calculate statistics for the examples in Chapter 6 of SDA, but you have already seen all of the features of SAS software used in that code.

Selecting a sample with unequal probabilities, however, involves features of the SURVEYSELECT procedure that we have not yet explored. Section 6.1 shows how to select a one-stage cluster sample with equal or unequal probabilities, and Section 6.2 shows how to select a two-stage sample.

6.1 Selecting a Sample with Unequal Probabilities

This section shows how to select a sample of primary sampling units (psus) with unequal probabilities using the SURVEYSELECT procedure. Subsampling all secondary sampling units (ssus) in the selected psus will give a one-stage cluster sample, and that includes the special case in which each cluster consists of one ssu—that is, an unequal-probability sample of elements.

6.1.1 Sampling with Replacement

Example 6.2 of SDA. In Chapters 2 through 5, samples were selected in the SURVEYSELECT procedure with METHOD=srs. Units, whether observation units or clusters, were selected with equal probabilities and without replacement within strata.

To select an unequal-probability sample with replacement, use METHOD=pps_wr. Since we are sampling with probability proportional to size (pps), we need to specify a SIZE variable that gives the relative sizes of the psus in the population data set *classes* (`classes.csv`). In Code 6.1, the SIZE variable is *class_size*, the number of students in the class.

Code 6.1. Select a with-replacement sample of 5 psus (`example0602.sas`).

```
PROC SURVEYSELECT DATA= classes OUT= mysample1 SAMPSIZE= 5 METHOD= pps_wr STATS
    SEED= 4082;
   SIZE class_size;

PROC PRINT DATA = mysample1;
RUN;
```

DOI: 10.1201/9781003160366-6

Output 6.1. Select a with-replacement sample of 5 psus (`example0602.sas`).

The SURVEYSELECT Procedure

Selection Method	PPS, With Replacement
Size Measure	class_size

Input Data Set	CLASSES
Random Number Seed	4082
Sample Size	5
Output Data Set	MYSAMPLE1

Obs	class	class_size	NumberHits	ExpectedHits	SamplingWeight
1	12	24	2	0.18547	5.39167
2	13	46	2	0.35549	2.81304
3	14	100	1	0.77280	1.29400

In Output 6.1, the *ExpectedHits* column is $n\psi_i$ (n times the draw-by-draw selection probability). For class 1, *ExpectedHits* $= (5)(0.068006) = 0.34003$. For class 14, *ExpectedHits* $= (5)(0.15456) = 0.7728$. The weights for with-replacement sampling are stored in variable *SamplingWeight* $= 1/$*ExpectedHits* $= 1/(n\psi_i)$.

Although I asked for a sample with SAMPSIZE=5, the printout of the data set *mysample1* has only three lines. Those three lines list the distinct units in the sample, and the variable *NumberHits* tells how many times each unit appears in the sample. Class 14, with *NumberHits* = 1, appears once in the sample. Each of classes 12 and 13 has *NumberHits* = 2, and each appears twice in the sample.

Each unit must be included *NumberHits* times in order to have a sample that produces approximately unbiased estimates of population quantities. If collecting data in two stages, you would take two independently selected subsamples from class 12 and two independently selected subsamples from class 13.

If you want to list the five units in the sample, call the SURVEYSELECT procedure with the OUTHITS option, as shown in Code 6.2.

Code 6.2. List repeated units from a with-replacement sample (`example0602.sas`).

```
PROC SURVEYSELECT DATA= classes OUT= mysample2 OUTHITS SAMPSIZE= 5 METHOD= pps_wr
    STATS SEED= 4082;
   SIZE class_size;

PROC PRINT DATA = mysample2;
RUN;
```

The first box in Output 6.2, from the SURVEYSELECT procedure, is the same as in Output 6.1 (except for the name of the data set containing the sample). The printed sample in data set *mysample2*, however, contains five observations, with class 12 listed twice and class 13 listed twice. When analyzing the data, make sure you use different psu names for the multiple instances of psus that appear more than once. For this example, you might want to use variable *psuid* with values 12.1, 12.2, 13.1, 13.2, and 14.

Output 6.2. List repeated units from a with-replacement sample (`example0602.sas`).

The SURVEYSELECT Procedure

Selection Method	PPS, With Replacement
Size Measure	class_size

Input Data Set	CLASSES
Random Number Seed	4082
Sample Size	5
Output Data Set	MYSAMPLE2

Obs	class	class_size	NumberHits	ExpectedHits	SamplingWeight
1	12	24	2	0.18547	5.39167
2	12	24	2	0.18547	5.39167
3	13	46	2	0.35549	2.81304
4	13	46	2	0.35549	2.81304
5	14	100	1	0.77280	1.29400

6.1.2 Sampling without Replacement

The SURVEYSELECT procedure offers many options for selecting a probability sample of units without replacement. Table 6.1 lists the methods discussed in SDA; additional methods that can be used to select an unequal-probability sample are described in the documentation for the SURVEYSELECT procedure (SAS Institute Inc., 2021). All of these methods can be used to select psus within strata by including the STRATA statement with the SURVEYSELECT procedure (see Section 7.1).

Example 6.10 of SDA. Code 6.3 selects a sample of size 15 from the population in `agpop.csv` with probability proportional to the number of acres devoted to farms in 1987. The variable *acres87*, however, has missing values for some of the counties, and the SURVEYSELECT procedure excludes any units with missing or nonpositive values for the SIZE variable from the sample selection. To ensure that all counties have a positive probability of selection, I define the variable *sizemeas* in the data set *agpop* to equal *acres87* if *acres87* > 0, *acres82* if *acres87* is missing and *acres82* > 0, and the median of the values of *acres87* otherwise.

Code 6.3. Select counties with unequal probabilities and without replacement (`example0610.sas`).

```
PROC SURVEYSELECT DATA= agpop METHOD= pps SAMPSIZE= 15 OUT= agpps JTPROBS SEED=
    783874;
  SIZE sizemeas;

PROC PRINT DATA = agpps;
  VAR county state sizemeas SelectionProb SamplingWeight Unit JtProb_1-JtProb_15;
RUN;
```

The code is similar to that for selecting the with-replacement sample in Code 6.2. The main difference is that METHOD=pps is requested instead of METHOD=pps_wr.

Code 6.3 requests the joint inclusion probabilities with the JTPROBS option. This option is needed only if you want to calculate the exact without-replacement variance, using the Horvitz-Thompson or Sen-Yates-Grundy formulas, for your survey. If you plan to calculate

TABLE 6.1
Options for selecting a probability sample referenced in Chapters 5 and 6 of SDA.

Method	Description
BREWER	Selects an unequal-probability sample without replacement containing 2 psus per stratum using Brewer's (1963, 1975) method. The SIZE variable contains the psu sizes, or, more generally, a quantity that is proportional to the desired inclusion probability π_i. The relative sizes should not exceed 1/2.
POISSON	Poisson sampling. The SIZE variable contains the desired inclusion probabilities π_i for each unit, and these should be between 0 and 1. Because the sample size is random, do not include SAMPSIZE= in the main PROC SURVEYSELECT statement. Also, do not include an ALLOC= option in the STRATA statement (if included) because the sample size in each stratum will be random.
PPS	Selects an unequal-probability sample without replacement using the Hanurav–Vijayan algorithm (Hanurav, 1967; Vijayan, 1968). The variable defined in the SIZE statement contains the population psu sizes M_i (or quantities proportional to the desired selection probabilities) for the units. This algorithm requires $nM_i \leq \sum_{j=1}^{N} M_j$ for all units; if your population does not meet this requirement either put the units with $M_i > \sum_{j=1}^{N} M_j/n$ in certainty strata or use the CERTSIZE option. This is my preferred method for selecting unequal-probability samples of psus, because it yields a sample with the desired inclusion probabilities, it gives joint inclusion probabilities that are strictly positive for all pairs of units, and the Sen-Yates-Grundy estimator of the variance is usually non-negative.
PPS_SYS	Selects an unequal-probability sample via systematic sampling. The variable defined in the SIZE statement contains the sizes (or quantities proportional to the desired selection probabilities π_i) for the units.
PPS_WR	Selects an unequal-probability sample with replacement. The SIZE variable contains the desired draw-by-draw selection probabilities, ψ_i (or a variable that is proportional to these probabilities).
SRS	Selects an SRS (or stratified random sample, if the STRATA statement is included) without replacement. Units are selected with equal probabilities that are determined by the sample sizes. Do not include a SIZE statement.
SYS	Selects a systematic sample. If you do not include a SIZE statement, this will select a systematic sample with equal probabilities. If you define SAMPSIZE= n, the procedure will select every kth unit, where $k = N/n$, using a random starting point. You can also define the sampling interval or starting point using options for the method. If you include a SIZE statement, METHOD=SYS is the same as METHOD=PPS_SYS.
URS	Selects a simple random sample (or stratified random sample, if the STRATA statement is included) with replacement. URS is an acronym for "unrestricted random sampling." Do not include a SIZE statement.

the with-replacement variance, as is usually recommended, you do not need to include the JTPROBS option. Output 6.3(a) shows the output from the SURVEYSELECT procedure, telling the selection method and size measure used to draw the sample.

Output 6.3(a). Select counties with unequal probabilities and without replacement (`example0610.sas`).

Selection Method	PPS, Without Replacement
Size Measure	sizemeas

Input Data Set	AGPOP
Random Number Seed	783874
Sample Size	15
Output Data Set	AGPPS

Output 6.3(b). Sample of counties with unequal probabilities and without replacement (`example0610.sas`).

Obs	county	state	sizemeas	SelectionProb	SamplingWeight	Unit	JtProb_1	JtProb_2	JtProb_3	JtProb_4	JtProb_5
1	HARRISON COUNTY	IN	184573	0.002865	349.076	1	0	.000007814	.000007983	.000008396	.000013025
2	ROCK ISLAND COUNTY	IL	187254	0.002906	344.078	2	.000007814	0	.000008099	.000008518	.000013215
3	WAYNE COUNTY	NY	191309	0.002969	336.785	3	.000007983	.000008099	0	.000008702	.000013501
4	ALLEN COUNTY	OH	201190	0.003123	320.244	4	.000008396	.000008518	.000008702	0	.000014198
5	TRAVERSE COUNTY	MN	312130	0.004844	206.420	5	.000013025	.000013215	.000013501	.000014198	0
6	HARDIN COUNTY	IA	337990	0.005246	190.627	6	.000014105	.000014310	.000014619	.000015375	.000023856
7	YOLO COUNTY	CA	505597	0.007847	127.433	7	.000021099	.000021406	.000021869	.000022999	.000035687
8	WILLIAMSON COUNTY	TX	527175	0.008182	122.217	8	.000021999	.000022319	.000022803	.000023981	.000037210
9	KIOWA COUNTY	CO	996785	0.015471	64.638	9	.000041597	.000042201	.000043115	.000045343	.000070356
10	WASCO COUNTY	OR	1172745	0.018202	54.939	10	.000048940	.000049651	.000050726	.000053347	.000082776
11	UMATILLA COUNTY	OR	1451108	0.022522	44.401	11	.000060556	.000061436	.000062767	.000066009	.000102423
12	WESTON COUNTY	WY	1546633	0.024005	41.658	12	.000064542	.000065480	.000066898	.000070355	.000109166
13	YAKIMA COUNTY	WA	1612399	0.025026	39.959	13	.000067287	.000068264	.000069743	.000073346	.000113808
14	CARTER COUNTY	MT	1629533	0.025292	39.539	14	.000068002	.000068990	.000070484	.000074126	.000115017
15	CUSTER COUNTY	MT	2183542	0.033890	29.507	15	.000091121	.000092445	.000094447	.000099327	.000154121

Obs	JtProb_6	JtProb_7	JtProb_8	JtProb_9	JtProb_10	JtProb_11	JtProb_12	JtProb_13	JtProb_14	JtProb_15
1	.000014105	.000021099	.000021999	.000041597	.000048940	.000060556	.000064542	.000067287	.000068002	.000091121
2	.000014310	.000021406	.000022319	.000042201	.000049651	.000061436	.000065480	.000068264	.000068990	.000092445
3	.000014619	.000021869	.000022803	.000043115	.000050726	.000062767	.000066898	.000069743	.000070484	.000094447
4	.000015375	.000022999	.000023981	.000045343	.000053347	.000066009	.000070355	.000073346	.000074126	.000099327
5	.000023856	.000035687	.000037210	.000070356	.000082776	.000102423	.000109166	.000113808	.000115017	.000154121
6	0	.000038645	.000040294	.000076188	.000089637	.000110914	.000118215	.000123242	.000124551	.000166896
7	.000038645	0	.000060294	.000114004	.000134129	.000165966	.000176891	.000184413	.000186373	.000249736
8	.000040294	.000060294	0	.000118875	.000139860	.000173057	.000184449	.000192293	.000194336	.000260406
9	.000076188	.000114004	.000118875	0	.000264834	.000327695	.000349267	.000364118	.000367988	.000493096
10	.000089637	.000134129	.000139860	.000264834	0	.000385836	.000411235	.000428721	.000433277	.000580583
11	.000110914	.000165966	.000173057	.000327695	.000385836	0	.000509595	.000531264	.000536910	.000719448
12	.000118215	.000176891	.000184449	.000349267	.000411235	.000509595	0	.000566595	.000572616	.000767294
13	.000123242	.000184413	.000192293	.000364118	.000428721	.000531264	.000566595	0	.000597249	.000800302
14	.000124551	.000186373	.000194336	.000367988	.000433277	.000536910	.000572616	.000597249	0	.000808912
15	.000166896	.000249736	.000260406	.000493096	.000580583	.000719448	.000767294	.000800302	.000808912	0

The SURVEYSELECT procedure creates variables *SelectionProb* (the inclusion probabilities), *SamplingWeight*, *Unit*, and *JtProb_1* through *JtProb_15*, shown in the sample printed in Output 6.3(b). Note that for each unit, $SelectionProb = 15\,(sizemeas)/966{,}449{,}249$, where 966,449,249 is the sum of *sizemeas* for all counties in the population, so that the probabilities are exactly proportional to *sizemeas*.

The variables *JtProb_1* through *JtProb_15* contain the joint inclusion probabilities requested by the JTPROBS option. These are indexed by the *Unit* variable. Thus, the entry in the first row and the *JtProb_7* column, 0.000021099, is the probability that units 1 and 7 are both in the sample, that is,

$$P\,(\text{Harrison County IN and Yolo County CA are both in the sample}) = 0.000021099.$$

As you can see, the matrix of the *JtProbs* will be large if you are selecting a large sample of psus. If that is the case, though, you will likely be calculating the with-replacement variance anyway and will not need the joint probabilities.

6.2 Selecting a Two-Stage Cluster Sample

Example 6.11 of SDA. A two-stage cluster sample is selected, no surprise, in two stages. First, select the psus, then select a sample of ssus from the selected psus. Here are the steps for selecting a two-stage cluster sample from the population of introductory statistics courses in data set `classes.csv`. The code for the full selection procedure is in file `select2stage.sas`.

Step 1. First, select the psus for the sample. Code 6.4 selects a probability-proportional-to-size sample of size 5 without replacement, similarly to Section 6.1. If you want to select psus with equal probabilities, use METHOD=srs instead of METHOD=pps. As always, you can set the SEED to any integer you like; this allows you to re-create the sample later.[1] If you want the procedure to generate the joint inclusion probabilities, include the JTPROBS option in the PROC SURVEYSELECT statement as was done in Code 6.3.

Code 6.4. Select the psus for a two-stage cluster sample (`select2stage.sas`).

```
PROC SURVEYSELECT DATA= classes OUT= psusample SAMPSIZE = 5 METHOD= pps SEED=
    75745;
   SIZE class_size; /* select psus with prob proportional to class_size */
   TITLE 'Draw sample of 5 psus with pps';

PROC PRINT DATA = psusample;
RUN;
```

Output 6.4(a). Select the psus for a two-stage cluster sample (`select2stage.sas`).

Draw sample of 5 psus with pps

The SURVEYSELECT Procedure

Selection Method	PPS, Without Replacement
Size Measure	class_size

Input Data Set	CLASSES
Random Number Seed	75745
Sample Size	5
Output Data Set	PSUSAMPLE

[1]Note that the sample in Output 6.4(b) has different psus than the sample in SDA. I selected the sample in SDA using an earlier release of SAS/STAT software. The software now uses the Mersenne-Twister random number generator (which has better statistical properties) to select samples, and in this book I wanted the example code to use the Mersenne-Twister algorithm. To obtain the sample of psus in Example 6.11 of SDA, run Code 6.4 with the RANUNI option (which calls the random number generator that was used in previous versions of SAS software) in the PROC SURVEYSELECT statement.

Output 6.4(b). Print the sample of psus (`select2stage.sas`).

Sample of 5 psus

Obs	class	class_size	SelectionProb	SamplingWeight
1	1	44	0.34003	2.94091
2	11	46	0.35549	2.81304
3	13	46	0.35549	2.81304
4	9	54	0.41731	2.39630
5	5	76	0.58733	1.70263

Step 2. Now that the sample of psus is selected, make a list of the ssus in each psu. For this small population, the DATA step of Code 6.5 creates variable *SsuID* that ranges from 1 to M_i. In other situations, you will need to obtain a list of ssus after selecting the sample of psus. For example, if taking a two-stage cluster sample of nursing home residents, you may have to find out the value of M_i, the number of residents, directly from each home and thus would know these values only after the first-stage sample is selected.

Code 6.5 also renames the variables containing the psu-level sampling weights and selection probabilities, which the SURVEYSELECT procedure creates in variables *SamplingWeight* and *SelectionProb*. A two-stage selection procedure has sampling weights and selection probabilities at both stages, and when we draw the ssus, the second-stage sampling weights and selection probabilities are again created in variables *SamplingWeight* and *SelectionProb*, overwriting the first-stage quantities that were previously created in those variables. Renaming the psu-level weight and probability variables as *psuweight* and *psuprob* retains the first-stage weights in variables that are not overwritten, and allows us to access them in later steps to calculate the final weights.

Code 6.5. List the ssus for psus selected in the two-stage cluster sample (`select2stage.sas`).

```
DATA onestage;
   SET psusample  (RENAME = (SamplingWeight = psuweight SelectionProb = psuprob) );
   DO SsuID = 1 TO class_size;
      OUTPUT;
   END;
RUN;
```

Step 3. Data set *onestage* contains a list of all ssus for the psus in the sample. If a one-stage cluster sample were desired, we would stop here. To obtain a two-stage cluster sample, select a probability sample from each of the psus in the sample. For this example, since the psus were selected with probability proportional to size, I want to take an SRS with the same number of ssus from each psu so that the sample of students is self-weighting.

In Code 6.6, I use the STRATA statement to select an SRS without replacement of 4 ssus from each psu. The SELECTALL option says to take all of the ssus in the psu if the desired sample size exceeds the population size M_i for that psu. The data set *onestage* contains only the psus selected at stage 1, and the STRATA statement provides a convenient way to select a subsample of ssus from each of those psus. This is still a cluster sample, though, and not a stratified sample because the sample of ssus is selected from the set of sampled psus.

Code 6.6. Select the ssus from each psu in the cluster sample (`select2stage.sas`).

```
PROC SURVEYSELECT DATA= onestage METHOD= srs SAMPSIZE= 4 OUT= SsuSample SEED= 34358
     STATS SELECTALL;
    STRATA class;
    TITLE 'Select subsample of 4 ssus within each of the sampled psus';
RUN;
```

Output 6.6. Select the ssus from each psu in the cluster sample (`select2stage.sas`).

Select subsample of 4 ssus within each of the sampled psus

The SURVEYSELECT Procedure

Selection Method	Simple Random Sampling
Strata Variable	class

Input Data Set	ONESTAGE
Random Number Seed	34358
Stratum Sample Size	4
Number of Strata	5
Total Sample Size	20
Output Data Set	SSUSAMPLE

Step 4. After Step 3, the data set *SsuSample* contains the sampled ssus from each sampled psu, along with the second-stage sampling weights and selection probabilities. I rename *SamplingWeight* and *SelectionProb* from the second-stage sample as *ssuweight* and *ssuprob* to avoid later confusion (otherwise, a future user of the data set might think *SamplingWeight* from the second-stage sample is the final weight). The final sampling weight *finalweight* for each unit in the sample is then computed in Code 6.7 as the product of the first-stage weight (in variable *psuweight*) and the second-stage weight (in variable *ssuweight*).

Code 6.7. Compute the final sampling weight (`select2stage.sas`).

```
DATA SsuSample;
   SET SsuSample (RENAME = (SamplingWeight = ssuweight SelectionProb = ssuprob) );
   finalweight = psuweight * ssuweight;

PROC PRINT DATA = SsuSample;
   VAR class class_size ssuid psuweight ssuweight finalweight;
   TITLE 'Cluster sample of ssus with sampling weights';
RUN;
```

And, since this is a small sample, let's take a look at it. Note that the final weights in Output 6.7 are the same for all ssus in the sample. The psus were selected with probabilities $5M_i/647$ and the ssus for psu i were selected with probability $4/M_i$, so the final weight is $[647/(5M_i)] \times [M_i/4] = 647/20 = 32.35$ for each ssu in the sample.

Output 6.7. Sample of ssus with final sampling weights.

Cluster sample of ssus with sampling weights

Obs	class	class_size	SsuID	psuweight	ssuweight	finalweight
1	1	44	8	2.94091	11.0	32.35
2	1	44	33	2.94091	11.0	32.35
3	1	44	37	2.94091	11.0	32.35
4	1	44	43	2.94091	11.0	32.35
5	5	76	24	1.70263	19.0	32.35
6	5	76	37	1.70263	19.0	32.35
7	5	76	42	1.70263	19.0	32.35
8	5	76	76	1.70263	19.0	32.35
9	9	54	9	2.39630	13.5	32.35
10	9	54	40	2.39630	13.5	32.35
11	9	54	49	2.39630	13.5	32.35
12	9	54	51	2.39630	13.5	32.35
13	11	46	18	2.81304	11.5	32.35
14	11	46	34	2.81304	11.5	32.35
15	11	46	43	2.81304	11.5	32.35
16	11	46	46	2.81304	11.5	32.35
17	13	46	22	2.81304	11.5	32.35
18	13	46	26	2.81304	11.5	32.35
19	13	46	40	2.81304	11.5	32.35
20	13	46	46	2.81304	11.5	32.35

Selecting a two-stage sample with unequal numbers of ssus. What if you want different sample sizes m_i in the psus? For example, if you select psus with equal probabilities, you may want to have larger second-stage sample sizes in psus with large values of M_i. Since we are treating the second stage of selection as drawing a stratified sample from the selected psus, the subsample sizes for psus can be specified using the custom allocation methods described in Section 3.1. Code 3.6 shows how to create a data set with the desired sample sizes that can be input into the SURVEYSELECT procedure. If it is desired to have m_i proportional to M_i, the second-stage sample can be selected with proportional allocation.

6.3 Computing Estimates from an Unequal-Probability Sample

The syntax used to compute estimates from an unequal-probability cluster sample is exactly the same as that used in Chapter 5 for equal-probability cluster samples. This section presents the code for three of the examples in Chapter 6 of SDA; code for the other examples is similar and can be found on the website.

The SURVEYMEANS and SURVEYFREQ procedures calculate estimates of means, totals, and proportions by using the formulas with the survey weights. When the TOTAL= option is omitted, standard errors are calculated with the formulas for the with-replacement variance in Section 6.4 of SDA.

When the TOTAL= option is included with a cluster sample, the SURVEYMEANS and SURVEYFREQ procedures multiply the with-replacement variance by $(1 - n/N)$, where n/N is the proportion of psus in the population that are included in the sample. When psus are selected with unequal probabilities, this gives neither the Horvitz-Thompson nor the Sen-Yates-Grundy without-replacement variance estimate discussed in Section 6.4 of SDA. The only situation in which the TOTAL= option gives the correct without-replacement variance is for a one-stage cluster sample that is selected with equal probabilities. In general, I recommend calculating the with-replacement variance estimate and omitting the TOTAL= option when unequal-probability sampling is used.

6.3.1 Estimates from With-Replacement Samples

Example 6.4 of SDA. Code 6.8 shows how to calculate estimates when the cluster total t_i has already been found for each psu (or when the psus are also the observation units, i.e., $M_i = 1$ for all psus). Since the summary statistic has already been calculated for each psu, no CLUSTER statement is needed. The only design feature that is specified in Code 6.8 is the WEIGHT statement. Class 14 appears twice in the data since it was selected twice for the sample—we call it class 141 for the first appearance and class 142 for the second to distinguish them.

The mean calculated from the SURVEYMEANS procedure estimates $\bar{t}_{\mathcal{U}}$, the population mean of the cluster totals t_i, which for this example is the total amount of time spent studying by students in class i. For this example, we want to estimate the average amount of time spent studying per student, which is the ratio $\hat{\bar{y}}_{\psi} = \hat{t}_{\psi}/\hat{M}_{\psi}$. The RATIO statement gives the estimate $\hat{\bar{y}}_{\psi}$. If the data set consists of the individual values y_{ij}, then the estimated mean from the SURVEYMEANS procedure will give the estimate $\hat{\bar{y}}_{\psi}$.

Code 6.8. One-stage pps cluster sampling (`example0604.sas`).

```
DATA studystat;
   INPUT class Mi tothours;
   classwt = 647/(Mi*5);
   datalines;
12 24 75
141 100 203
142 100 203
5 76 191
1 44 168
;

PROC SURVEYMEANS DATA = studystat NOBS MEAN SUM CLM CLSUM;
   WEIGHT classwt;
   VAR tothours;
   RATIO tothours/Mi;
RUN;
```

Output 6.8. One-stage pps cluster sampling (`example0604.sas`).

Statistics									
Variable	N	Mean	Std Error of Mean	95% CL for Mean		Sum	Std Error of Sum	95% CL for Sum	
tothours	5	138.555465	33.935346	44.3358394	232.775090	1749.014359	222.421820	1131.47239	2366.55633
Mi	5	51.254803	15.789650	7.4157059	95.093899	647.000000	0	647.00000	647.00000

Ratio Analysis						
Numerator	Denominator	N	Ratio	Std Error	95% CL for Ratio	
tothours	Mi	5	2.703268	0.343774	1.74879812	3.65773776

Example 6.6 of SDA. The estimates for a two-stage cluster sample with replacement are calculated exactly the same way as a one-stage sample. Code 6.9 shows the DATA step. For this example, we have data for the individual students in the psus so we enter those for each student.

Code 6.9. Two-stage pps cluster sampling with replacement (`example0606.sas`).

```
DATA students;
   INPUT class popMi sampmi hours;
   studentwt = (647/(popMi*5)) * (popMi/sampmi);
   DATALINES;
12    24 5 2
12    24 5 3
12    24 5 2.5
12    24 5 3
12    24 5 1.5
141 100 5 2.5
141 100 5 2
141 100 5 3
141 100 5 0
141 100 5 0.5
142 100 5 3
142 100 5 0.5
142 100 5 1.5
142 100 5 2
142 100 5 3
5     76 5 1
5     76 5 2.5
5     76 5 3
5     76 5 5
5     76 5 2.5
1     44 5 4
1     44 5 4.5
1     44 5 3
1     44 5 2
1     44 5 5
;

PROC SURVEYMEANS DATA = students DF MEAN SUM CLM CLSUM;
   WEIGHT studentwt;
   CLUSTER class;
   VAR hours;
RUN;
```

Class 14 appears twice in the sample of psus in Example 6.6. An independent set of students is selected for each appearance. To enable correct variance calculations, the first occurrence of class 14 is relabeled as class 141, and the second occurrence as class 142. These are counted as two separate psus in the estimation. If you labeled both as 14, the SURVEYMEANS procedure would treat that as one psu with 10 ssus instead of two psus with 5 ssus each.

The weight *studentwt* is calculated as the first-stage weight $(\sum_{j=1}^{N} M_j)/(nM_i)$ times the second-stage weight M_i/m_i. In most samples, there is some deviation from a strict pps design, so it is good practice to calculate the weights directly, using the selection probabilities. Here, the weight for each student simplifies to 647/25 because the sample is self-weighting.

For many problems, defining the weights is the trickiest part, and it is also the most important. Always check that the sum of the weights approximately (or exactly for this example, as seen in the Data Summary of Output 6.9) equals the population size.

Output 6.9. Two-stage pps cluster sampling with replacement (`example0606.sas`).

The SURVEYMEANS Procedure

Data Summary	
Number of Clusters	5
Number of Observations	25
Sum of Weights	647

Statistics

Variable	DF	Mean	Std Error of Mean	95% CL for Mean		Sum	Std Error of Sum	95% CL for Sum	
hours	4	2.500000	0.360555	1.49893848	3.50106152	1617.500000	233.279168	969.813197	2265.18680

The estimated population total and mean are the same values as in Example 6.6 of SDA. Note that 4 degrees of freedom (1 less than the number of psus) are used for the confidence interval.

6.3.2 Estimates from Without-Replacement Samples

Estimates from without-replacement cluster samples are calculated exactly the same way as estimates from with-replacement samples. Use the WEIGHT statement to provide the sampling weights at the observation-unit level, and use the CLUSTER statement to provide the information on psu membership. This gives the with-replacement variance, which is usually only slightly larger than the without-replacement variance. The formulas for the Sen-Yates-Grundy variance estimate must be used if it is desired to estimate the without-replacement variance.

Example 6.11 of SDA. Code 6.10, analyzing a without-replacement pps sample, has the same form as Code 6.9. Even though the sample was selected without replacement, the with-replacement variance is calculated for simplicity and stability.

Code 6.10. Two-stage pps cluster sample without replacement (`example0611.sas`).

```
PROC SURVEYMEANS DATA = classpps DF MEAN SUM CLM CLSUM;
   WEIGHT finalweight;
   CLUSTER class;
   VAR hours;
RUN;
```

Output 6.10. Two-stage pps cluster sample without replacement (`example0611.sas`).

Data Summary	
Number of Clusters	5
Number of Observations	20
Sum of Weights	647

Statistics

Variable	DF	Mean	Std Error of Mean	95% CL for Mean		Sum	Std Error of Sum	95% CL for Sum	
hours	4	3.450000	0.481858	2.11214665	4.78785335	2232.150000	311.762373	1366.55889	3097.74111

6.4 Summary, Tips, and Warnings

The SURVEYSELECT procedure can be used to select equal-probability and unequal-probability cluster samples. The general form of the statement for selecting an unequal-probability cluster sample is given below. Table 6.1 lists some of the METHODs that can be used to select samples.

```
PROC SURVEYSELECT DATA = population_data OUT = sample_data SAMPSIZE = sample_size
    METHOD = pps_method STATS SEED = seed_value;
   SIZE size_variable;
```

The syntax of the SURVEYMEANS and SURVEYFREQ procedures for analyzing data from a cluster sample, given in Section 5.5, is the same for both equal-probability and unequal-probability samples.

Tips and Warnings

- When selecting a probability-proportional-to-size sample with the SURVEYSELECT procedure, check the output values of the selection probabilities to make sure these are roughly proportional to the unit sizes.
- The more complex the sampling plan, the more complicated the weight calculations. Always check that the sum of the weights given by the survey analysis procedure approximately equals the population size.
- For unequal-probability sampling, omitting the TOTAL= option in the survey analysis procedures gives the with-replacement variance. This is the approach that I recommend for most surveys. If the without-replacement variance is desired, use the Sen-Yates-Grundy formula directly.
- If a with-replacement cluster sample is drawn, give psus that are drawn more than once a separate identifier for each appearance in the sample.

7

Complex Surveys

We have already seen most of the components needed for selecting and computing estimates from a stratified multistage sample. Now let's put them all together. Section 7.1 summarizes the steps for selecting a stratified multistage sample, and Section 7.3 tells how to compute estimates, using data from the National Health and Nutrition Examination Survey (NHANES).

The new features considered in this chapter are how to estimate quantiles (Section 7.2) and how to graph survey data (Sections 7.4 and 7.5).

7.1 Selecting a Stratified Multistage Sample

A stratified multistage sample is selected in stages, and we have already seen how to do all of the steps. Let's review them here.

Select the primary sampling units (psus). Code 7.1 selects a sample of classes from the small population considered in Section 6.2 of SDA, after first dividing the classes into three strata based on their sizes. Stratum 1 contains the two large classes, stratum 2 contains six medium-sized classes, and stratum 3 contains the seven smallest classes. The code specifies drawing two psus with probability proportional to *class_size* from each stratum. The psus are arranged in strata, so the only new feature here is adding the STRATA statement to the SURVEYSELECT procedure used to select the psus.

Code 7.1. Select 2 psus from each stratum (`select_stratmulti.sas`).

```
DATA classes;
   SET datalib.classes;
   IF class_size > 70 THEN strat = 1;
   ELSE IF class_size > 40 THEN strat = 2;
   ELSE strat = 3;

PROC SURVEYSELECT DATA = classes OUT = psusample SAMPSIZE = 2 METHOD = pps STATS
     SEED = 187263;
   STRATA strat;
   SIZE class_size;

PROC PRINT DATA = psusample;
RUN;
```

Output 7.1 illustrates a two-psu-per-stratum design, which is commonly used for stratified multistage samples because it has a high degree of stratification but still allows variance

DOI: 10.1201/9781003160366-7

estimates to be calculated within each stratum. Of course, other allocations can be used for numbers of psus to be sampled from strata, as described in Chapter 3.

Output 7.1(a) shows the output generated by the SURVEYSELECT procedure, and Output 7.1(b) shows the list of psus selected for the sample. Stratum 1 contains two psus, so the procedure selects each of these with certainty and assigns them *SamplingWeight* one.

Output 7.1(a). SURVEYSELECT procedure output (`select_stratmulti.sas`).

The SURVEYSELECT Procedure

Selection Method	PPS, Without Replacement
Size Measure	class_size
Strata Variable	strat

Input Data Set	CLASSES
Random Number Seed	187263
Stratum Sample Size	2
Number of Strata	3
Total Sample Size	6
Output Data Set	PSUSAMPLE

Output 7.1(b). Psus selected for stratified multistage sample (`select_stratmulti.sas`).

Obs	strat	class	class_size	SelectionProb	SamplingWeight
1	1	5	76	1.00000	1.00000
2	1	14	100	1.00000	1.00000
3	2	8	44	0.29630	3.37500
4	2	9	54	0.36364	2.75000
5	3	12	24	0.27586	3.62500
6	3	2	33	0.37931	2.63636

After the psus are selected for the sample, secondary sampling units, and units in subsequent stages of sampling, are selected as described in Steps 2 through 4 of Section 6.2.

7.2 Estimating Quantiles

Quantiles are estimated using the empirical cumulative distribution function (cdf) $\hat{F}(y)$, which is the sum of the weights for sample observations having $y_i \leq y$ divided by the sum of all weights for the sample. Because $\hat{F}(y)$ has jumps at the distinct values of y in the sample, however, for many values of q there is no value of y in the sample that has $\hat{F}(y)$ exactly equal to q. Multiple definitions for population and sample quantiles have been proposed (Hyndman and Fan, 1996; Wang, 2021).

The SURVEYMEANS procedure calculates quantiles using an interpolated empirical cdf, discussed in Exercise 7.19 of SDA. Let $y_{(1)} < y_{(2)} < \ldots < y_{(K)}$ be the distinct values of y in the sample (these are the values at which $\hat{F}$ jumps, ordered from smallest to largest). The

qth quantile (also called the 100qth percentile) is calculated as

$$\tilde{\theta}_q = \begin{cases} y_{(1)} & \text{if } q < \hat{F}(y_{(1)}) \\ y_{(k)} + \dfrac{q - \hat{F}(y_{(k)})}{\hat{F}(y_{(k+1)}) - \hat{F}(y_{(k)})}[y_{(k+1)} - y_{(k)}] & \text{if } \hat{F}(y_{(k)}) \le q < \hat{F}(y_{(k+1)}) \\ y_{(K)} & \text{if } q = 1 \end{cases} \tag{7.1}$$

The interpolated quantiles $\tilde{\theta}_q$ are calculated by requesting the desired quantile values in the statistics keywords of the SURVEYMEANS statement. Request the 0.25, 0.5, 0.75, and 0.90 quantiles, for example, by typing QUANTILE=(.25 .50 .75 .90). You can, equivalently, request percentiles by typing PERCENTILE=(25 50 75 90). The calculations are exactly the same with either keyword; the only difference is that requested quantiles are between 0 and 1 (probabilities), and requested percentiles are between 0 and 100 (percents).

Example 7.5 of SDA. Code 7.2 requests quantiles for the simple random sample (SRS) of height values in file `htsrs.csv`. Of course, you would also include the STRATA and CLUSTER statements for a survey with those design features, as we shall see in Code 7.4.

Code 7.2. Calculate interpolated quantiles from the SRS of heights (`example0705.sas`).

```
PROC SURVEYMEANS DATA = htsrs QUANTILE = (.25 .50 .75 .90);
   WEIGHT wt;
   VAR height;
RUN;
```

Output 7.2. Calculate interpolated quantiles from the SRS of heights (`example0705.sas`).

Data Summary	
Number of Observations	200
Sum of Weights	2000

Quantiles						
Variable	Percentile		Estimate	Std Error	95% Confidence Limits	
height	25	Q1	159.700000	0.875361	157.973827	161.426173
	50	Median	168.750000	0.974641	166.828051	170.671949
	75	Q3	176.000000	1.094346	173.841997	178.158003
	90	D9	183.400000	1.504112	180.433956	186.366044

7.3 Computing Estimates from Stratified Multistage Samples

We have seen all the building blocks for computing the estimates from any survey. Now let's put them all together using the data from the National Health and Nutrition Examination Survey (NHANES). The Centers for Disease Control and Prevention produce online tutorials for analyzing NHANES data. These can be found at `https://wwwn.cdc.gov/nchs/nhanes/tutorials/`, with tips for analysis in SAS software at `https://wwwn.cdc.gov/nchs/nhanes/tutorials/softwaretips.aspx`.

Example 7.9 of SDA looks at statistics about body mass index (BMI, variable *bmxbmi*) for adults age 20 and over. Let's compute those estimates in the SURVEYMEANS procedure.

First, note that the subset of adults age 20 and over is a domain for the data, so as a first step, Code 7.3 defines the domains using the age variable *ridageyr* so that we can produce statistics for the domain with *age20*=1, as described in Section 4.3. It also defines the variable *bmi30* to equal 1 if the person's BMI is greater than 30 and 0 if it is less than or equal to 30. Note that the domain-defining variable *age20* is available for every record in the data set (there are no missing values), but *bmi30* is missing if variable *bmxbmi* is missing in the data.

Code 7.3. Define the domains for persons under and over age 20 (`example0709.sas`).

```
DATA nhanes;
   SET datalib.nhanes;
   IF ridageyr >= 20 THEN age20 = 1;
   ELSE age20 = 0;
   bmi30 = .;
   IF bmxbmi > 30 THEN bmi30 = 1;
   IF 0 < bmxbmi <= 30 THEN bmi30 = 0;
RUN;
```

Code 7.4 gives the commands for the SURVEYMEANS procedure to calculate the estimates. For now let's suppress the plots (PLOTS=NONE); we'll come back to graphing the data in Section 7.4.

Code 7.4. Calculate BMI statistics for the domain of adults (`example0709.sas`).

```
PROC SURVEYMEANS DATA = nhanes PLOTS = NONE DF NOMCAR MEAN CLM QUANTILE = (0.05
    0.25 0.5 0.75 0.95);
  CLASS bmi30;
  WEIGHT wtmec2yr;
  STRATA sdmvstra;
  CLUSTER sdmvpsu;
  DOMAIN age20;
  VAR bmxbmi bmi30;
RUN;
```

You have already seen almost all of the components in Code 7.4. As always, the WEIGHT statement specifies the final weight for the observation units. The file `nhanes.csv` contains two weight variables: *wtint2yr* gives the weight for the set of persons with interview data, and *wtmec2yr* gives the weight for the subset of interviewed persons who had a medical examination. BMI is measured in the medical examination, so the appropriate weight variable to use is *wtmec2yr*. Variable *bmi30* is defined as a categorical variable in the CLASS statement so that the procedure will estimate the proportions in each category; because *bmi30* takes on values 0 and 1, it would also work to omit the CLASS statement and estimate the mean of *bmi30*.

The STRATA and CLUSTER statements are used exactly as in Chapters 3, 5, and 6, except now both are used together to indicate that the psus are nested within strata. The combination of the STRATA and CLUSTER variables must uniquely identify each psu. For the NHANES data, *sdmvstra* runs from 119 to 133 and *sdmvpsu* identifies the psu (1 or 2) within each stratum. The TOTAL= option is omitted from the PROC SURVEYMEANS statement in Code 7.4. With complex samples such as the NHANES data, we usually want to calculate the with-replacement variance, which requires only psu-level information.

The DOMAIN statement, too, is an old friend from Chapter 4. This specifies that separate estimates are desired for the two domains defined by *age20*, and the procedure calculates the estimates specified by the statistics keywords for each domain.

Code 7.4 has an additional option in the first line that we have not seen before. There are missing data in NHANES, and in general, the missing data cannot be treated as though they are missing completely at random (MCAR), as defined in Section 8.4 of SDA. The NOMCAR option in the SURVEYMEANS procedure tells it to treat the observations that are missing the response variable *bmxbmi* as a domain (where the sample size is a random variable) so that standard errors are correct if one or more psus are missing that variable for all observations.

The first output box from the procedure (not shown here) produces statistics for everyone in the data set. But we are interested only in the domain for adults age 20 and over (having *age20* = 1). The domain statistics output in Output 7.4 gives the statistics for both domains. The statistics for adults are in the rows corresponding to *age20* = 1.

The top box in Output 7.4 gives the estimated mean of BMI and the proportion in each category of *bmi30*. The mean BMI for adults age 20 and over is 29.389, with 95% confidence interval [28.849, 29.929]. The estimated proportion of adults age 20 and over who have BMI $>$ 30 is 0.392225 with 95% confidence interval [0.3584293, 0.4260202]. The confidence intervals are calculated using a t distribution with 15 (number of psus minus number of strata) degrees of freedom.

Output 7.4. Calculate BMI statistics for the domain of adults (`example0709.sas`).

Statistics for age20 Domains

age20	Variable	Level	Label	DF	Mean	Std Error of Mean	95% CL for Mean	
0	bmxbmi		Body Mass Index (kg/m**2)	15	20.602669	0.195750	20.1854377	21.0198996
	bmi30	0		15	0.930032	0.008746	0.9113898	0.9486739
		1		15	0.069968	0.008746	0.0513261	0.0886102
1	bmxbmi		Body Mass Index (kg/m**2)	15	29.389101	0.253197	28.8494243	29.9287768
	bmi30	0		15	0.607775	0.015856	0.5739798	0.6415707
		1		15	0.392225	0.015856	0.3584293	0.4260202

Quantiles for age20 Domains

age20	Variable	Label	Percentile		Estimate	Std Error	95% Confidence Limits	
0	bmxbmi	Body Mass Index (kg/m**2)	5		14.555706	0.078170	14.3890899	14.7223219
			25	Q1	16.244182	0.096211	16.0391121	16.4492514
			50	Median	18.988596	0.212578	18.5354975	19.4416955
			75	Q3	23.244709	0.305510	22.5935301	23.8958872
			95		31.943917	0.786276	30.2680101	33.6198242
1	bmxbmi	Body Mass Index (kg/m**2)	5		20.298934	0.180043	19.9151811	20.6826869
			25	Q1	24.353490	0.214092	23.8971646	24.8098151
			50	Median	28.234898	0.317489	27.5581854	28.9116100
			75	Q3	33.066151	0.307886	32.4099066	33.7223956
			95		42.640916	0.353799	41.8868113	43.3950198

The second box in Output 7.4 gives the quantiles for numeric variable *bmxbmi*; the procedure does not compute quantiles for CLASS variables. The median BMI for adults age 20 and over is 28.235, with 95% confidence interval [27.558, 28.912].

All of the features of the SURVEYMEANS procedure that were discussed in earlier chapters apply to stratified multistage samples. You can calculate ratios, for example, by including the RATIO statement. We'll return to poststratification in Chapter 8.

Using the SURVEYFREQ procedure with complex survey data. We can also calculate the percentage of adults who have BMI > 30 (variable *bmi30*) using the SURVEYFREQ procedure. The point estimate and standard error will be the same as the calculations using Code 7.4. The SURVEYFREQ procedure will also calculate design effects (deffs) for estimated proportions. (We'll see how to calculate deffs for numeric variables in Section 11.2.2.)

The SURVEYFREQ procedure does not have a DOMAIN statement. To calculate percentages and standard errors for statistics calculated within a domain, request a cross-classified table where the first dimension is the domain variable (here, *age20*).

When two variables are cross-classified, the TABLES statement treats the first variable listed as the row variable:

```
TABLES row_variable * column_variable;
```

The TABLES statement in Code 7.5 requests a cross-classification of the counts for the variables *age20* and *bmi30*, with *age20* serving as the row variable. The ROW option then requests that percentages of each category of *bmi30* be calculated for each level of the row variable in the table, *age20*. Adding (DEFF) behind the ROW option requests that deffs be computed for each row statistic.

Code 7.5. Estimate domain percentages and deffs for *bmi30* (`example0709.sas`).

```
PROC SURVEYFREQ DATA = nhanes NOMCAR;
  WEIGHT wtmec2yr;
  STRATA sdmvstra;
  CLUSTER sdmvpsu;
  TABLES age20*bmi30/ CL ROW(DEFF);
RUN;
```

Output 7.5 contains all the information we want, but it takes a bit of work to dig it out.

The row in Output 7.5 corresponding to *age20* $= 1$ and *bmi30* $= 1$ contains the percentage of adults age 20 and over who have BMI > 30. That percentage is given by the "Row Percent" 39.2225, with 95% confidence interval [35.8429, 42.6020]. The estimated design effect for this statistic, the ratio of the variance of the estimated percentage under the complex survey design to $\hat{p}(1-\hat{p})/n$, is 5.7001.

Output 7.5. Estimate domain percentages and deffs for *bmi30* (`example0709.sas`).

Table of age20 by bmi30

age20	bmi30	Frequency	Weighted Frequency	Std Err of Wgt Freq	Percent	Std Err of Percent	95% Confidence Limits for Percent		Row Percent	Std Err of Row Percent	95% Confidence Limits for Row Percent		Design Effect of Row Percent
0	0	3130	68019100	5213406	22.3070	0.8949	20.3997	24.2144	93.0032	0.8746	91.1390	94.8674	3.9369
	1	220	5117213	606982	1.6782	0.2316	1.1845	2.1719	6.9968	0.8746	5.1326	8.8610	3.9369
	Total	3350	73136313	5310566	23.9852	0.9754	21.9062	26.0643	100.0000				
1	0	3248	140873706	8737922	46.1999	1.4759	43.0541	49.3457	60.7775	1.5856	57.3980	64.1571	5.7001
	1	2158	90912164	5877609	29.8149	1.1652	27.3314	32.2983	39.2225	1.5856	35.8429	42.6020	5.7001
	Total	5406	231785870	12531715	76.0148	0.9754	73.9357	78.0938	100.0000				
Total	0	6378	208892806	12559128	68.5069	1.2245	65.8970	71.1168					
	1	2378	96029377	5886628	31.4931	1.2245	28.8832	34.1030					
	Total	8756	304922183	16602971	100.0000								

Frequency Missing = 788

If you want to obtain deffs for the cross-classified percentages in addition to deffs for the row percentages, use options CL DEFF ROW(DEFF) in the TABLES statement.

7.4 Univariate Plots from Complex Surveys

The SURVEYMEANS procedure produces histograms, boxplots, and density estimates of survey data. The DOMAIN statement produces side-by-side boxplots for domains. These plots are all calculated using the survey weights, as described in Chapter 7 of SDA.

Examples 7.10, 7.11, and 7.12 of SDA. These examples consider data in *htstrat*, a disproportional stratified sample of 160 women and 40 men. Calling the SURVEYMEANS procedure with the PLOTS=ALL option (see Code 7.6) creates a histogram and boxplot of the full data on the same graph.

Two density estimates are produced: one is a kernel density estimate where the bandwidth (discussed in Example 7.12 of SDA) is chosen automatically by the program, and the other is the density from a normal distribution with the mean and standard deviation estimated from the data.

Code 7.6. Histogram and boxplots for stratified sample of heights (`example0710.sas`).

```
PROC SURVEYMEANS DATA = htstrat PLOTS = ALL;
   WEIGHT wt;
   STRATA gender;
   VAR height;
   DOMAIN gender;
RUN;
```

Output 7.6(a). Histogram and boxplot for all data in stratified sample of heights (`example0710.sas`).

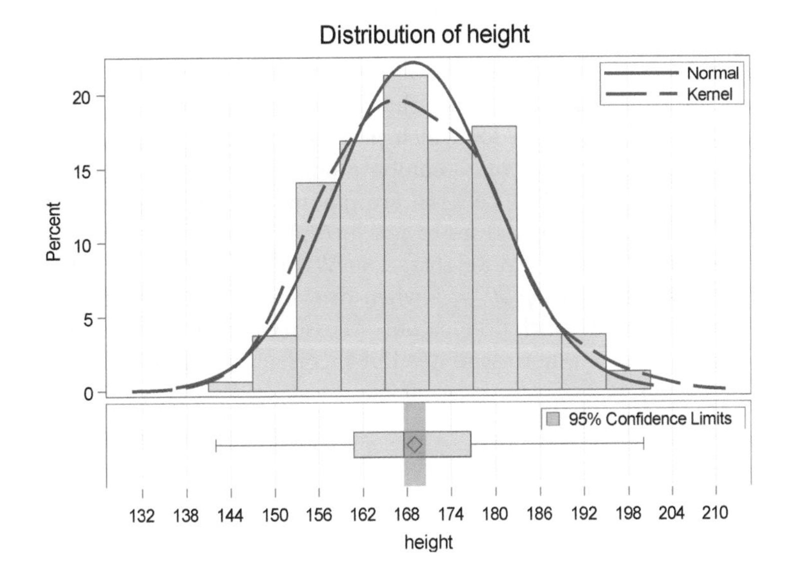

Output 7.6(a) shows the plots for the full data set with the default settings. The SURVEYMEANS procedure allows some customization of the plots. For example, you can plot the histogram and boxplot separately, and control the number of bins in the histogram with the NBINS option; these variations are shown in the code `example0710.sas` on the book website. For a higher degree of customization, you can also construct graphs by computing the appropriate quantities with weights and then writing commands to draw graphs with the SGPLOT procedure. Lohr (2012) provides code for using SAS/GRAPH® procedures to produce univariate plots and scatterplots from complex survey data.

The DOMAIN statement in Code 7.6, along with the PLOTS=ALL option in the SURVEYMEANS statement, creates a boxplot showing the quantiles for the full data and each domain. This is shown in Output 7.6(b).

Output 7.6(b). Draw side-by-side boxplots for domains from stratified sample of heights (`example0710.sas`).

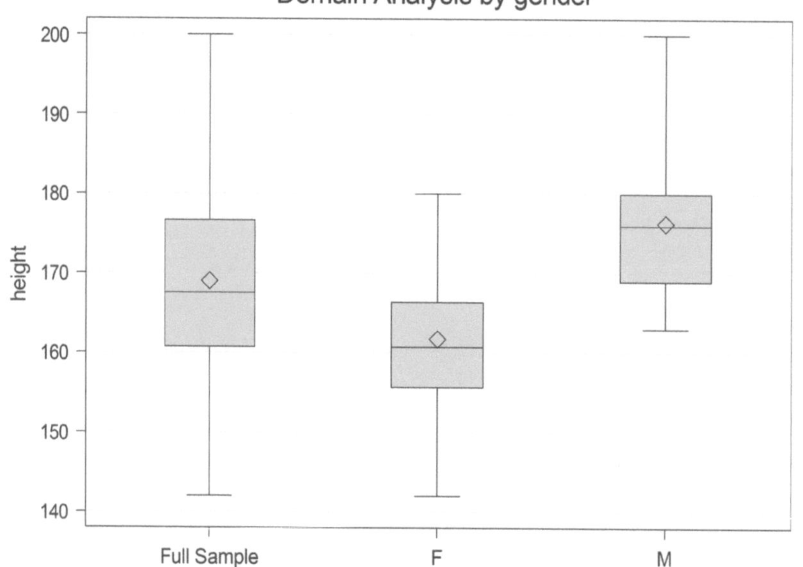

Histograms for domains. What if you want to draw a histogram for just one domain from a complex survey? You can do that by using the PLOTS option in the SURVEYMEANS procedure with the subset of the data for which plots are desired. Yes, I know I said that you should use the DOMAIN statement to compute domain statistics instead of conducting analyses on subsets the data, but here we are not producing standard errors: we are *only* using the SURVEYMEANS procedure to produce plots that estimate the population histogram by incorporating the survey weights. The WHERE statement in Code 7.7 says to use only the observations with $age20 = 1$ when drawing the histogram. The STRATA and CLUSTER statements are omitted since only the survey weights are needed to draw the plot: Code 7.7 gives correct point estimates but incorrect standard errors.

Code 7.7. Histogram of body mass index for adults age 20+ (`nhanesgraphs.sas`).

```
/* This code gives correct point estimates and graphs for the domain, but the
    standard errors are wrong because the data are subsetted and no STRATA and
    CLUSTER statements are used. Use this code ONLY for drawing graphs of a domain.
     */
PROC SURVEYMEANS DATA = nhanes PLOTS = (HISTOGRAM(NBINS=30));
  WHERE age20 = 1;
  WEIGHT wtmec2yr;
  VAR bmxbmi;
RUN;
```

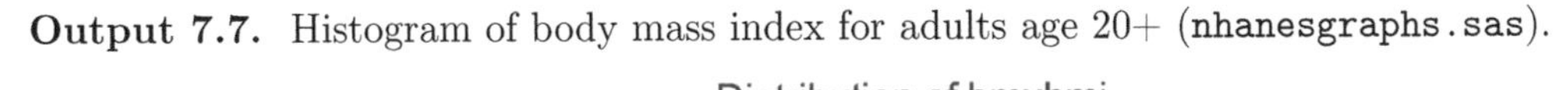

Output 7.7. Histogram of body mass index for adults age 20+ (`nhanesgraphs.sas`).

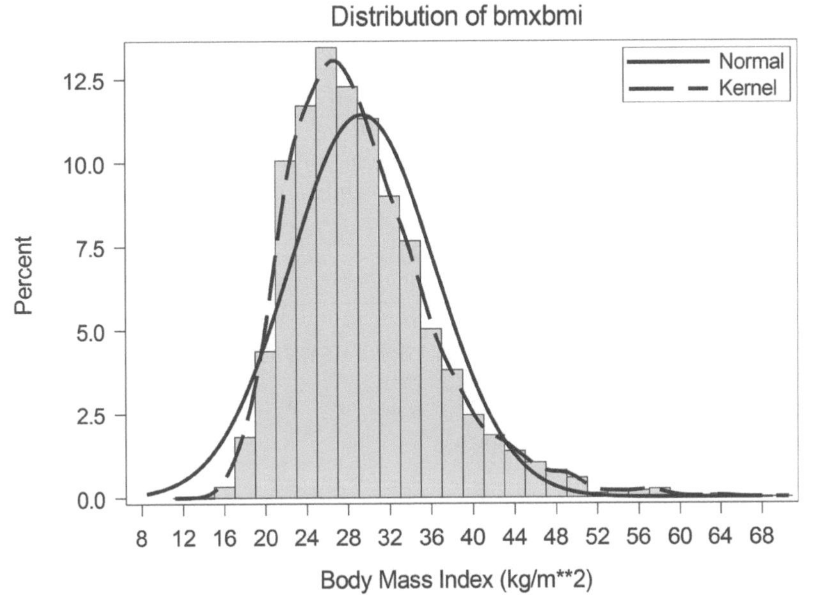

Output 7.7 shows the smoothed kernel density estimate and the best-fitting normal density. It is clear from the graph that the distribution is skewed. The superimposed normal density does not fit the data well.

7.5 Scatterplots from Complex Surveys

All of the plot types described in Section 7.6 of SDA can be drawn with SAS software. Some of those plots are shown in this section; the code in `nhanesgraphs.sas` shows how to construct additional plots. Some of the plots can be drawn by requesting the PLOTS= options in the SURVEYMEANS and SURVEYREG procedures. Others can be drawn using the SGPLOT procedure, which draws a variety of different plot types.

Warning: Remember that the survey weights can be used to produce any point estimate from survey data, but information about the stratification and clustering is needed to calculate standard errors. Some of the graph calculations below make use of procedures, such as the MEANS and QUANTREG procedures, that incorporate weights into the calculations but do not consider the stratification and clustering. Output from the code in this section should be used *only* for creating graphs. If you want to report statistics with correct standard errors, you must use one of the survey analysis procedures or perform special calculations.

Unweighted plots. The SGPLOT procedure draws histograms, boxplots, scatterplots, and more. Most of the plot types do not incorporate the survey weights. In general, extra data preparation is needed to persuade the SGPLOT procedure to draw graphs that estimate the population.

In some instances, however, you may want to examine the unweighted data. You may want to see how the weights affect the regression relationship between x and y, or to identify unusual observations in the data. For that reason, and to introduce you to the SGPLOT procedure, I include an unweighted scatterplot of the NHANES data.

Code 7.8 shows the code to produce an unweighted scatterplot of y variable body mass index (*bmxbmi*) versus x variable age (*ridageyr*). The SCATTER statement requests a scatterplot with the designated X and Y variables. You do not need to include the slash (/) in the SCATTER statement or any of the material behind it, but, if desired, numerous options are available for customizing the plot; I chose to use a black '+' symbol for the points. The optional XAXIS and YAXIS statements allow customizing the axis labels, ranges, and other features. I set the minimum and maximum values for each axis so that all plots in this section will have the same scale.

Code 7.8. Plot with no weights (`nhanesgraphs.sas`).

```
PROC SGPLOT DATA = nhanes;
  SCATTER X = ridageyr Y = bmxbmi / MARKERATTRS = (SYMBOL=Plus COLOR=black);
  XAXIS LABEL = "Age (years)" MIN = 0 MAX = 80;
  YAXIS LABEL = "Body Mass Index" MIN = 10 MAX = 70;
  TITLE 'Scatterplot of data---No weights';
RUN;
```

Output 7.8. Plot with no weights (`nhanesgraphs.sas`).

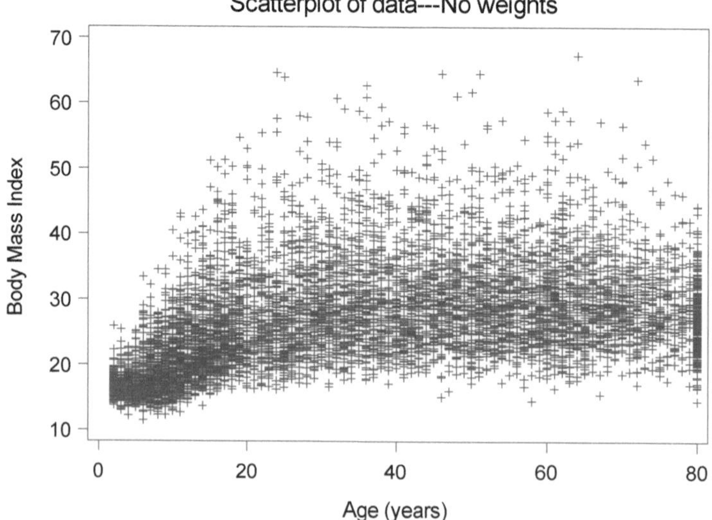

The unweighted scatterplot shows the relationship between x and y in the sample. If the sample is self-weighting, this scatterplot also estimates the relationship between x and y in the population. If the sample is not self-weighting, however, the relationship between x and y in the population may differ from the relationship in the sample, and one of the other plots in this section should be used.

Although I do not recommend an unweighted plot for non-self-weighting samples in general, sometimes, for small data sets, it is useful for seeing whether the relationship between x and y is the same with and without the weights (see Section 11.4 of SDA). The NHANES data, however, have so many data points that it is difficult to see any patterns at all in Output 7.8. The other plots in this section are better choices for the NHANES data.

Bubble plots. A bubble plot uses the survey weights to display the estimated relationship between x and y in the population. It can thus be used with samples that are not self-weighting as well as those that are. Bubble plots can be constructed using individual (x, y) values or using bins. If the data have only a few distinct values of (x, y) pairs—for example, when each variable is on a scale from 1 to 5—then the plot should be constructed to display all distinct (x, y) pairs. If x or y take on many distinct values—for example, when y is height measured to multiple decimal places—then a binned plot should be drawn.

To create bubble plots where the bubble areas are proportional to the weights at individual (x, y) values, first sum the weights at each distinct combination of x and y, as in Code 7.9, so that the bubble area will reflect all of the sample data having those values of x and y. If you skip this step, the SGPLOT procedure will attempt to display separate bubbles for each observation having a particular (x, y) value, and some of those bubbles may be hidden.[1] If the data set has a large number of distinct (x, y) pairs, make sure you use the NOPRINT option in the MEANS procedure so that you do not generate a long string of output in the Results window (if you forget to include the NOPRINT option, you will see what I mean).

Code 7.9. Bubble plot: Find the sum of the weights at each distinct (x, y) pair (`nhanesgraphs.sas`).

```
/* First, sort the data by the x and y variables */
PROC SORT DATA = nhanes;
   BY ridageyr bmxbmi;

/* Find the weight sum at each distinct (x,y) pair */
PROC MEANS DATA = nhanes NOPRINT SUM;
   BY ridageyr bmxbmi;
   VAR wtmec2yr;
   OUTPUT OUT = bubbleage  SUM = sumwts;
RUN;
```

The OUTPUT statement in Code 7.9 creates the data set *bubbleage*, containing variables *ridageyr* (x), *bmxbmi* (y), and *sumwts* (the sum of the weights for all observations having those values of x and y). Now that the data set has been created, you can use the SGPLOT procedure to draw the bubble plot, as shown in Code 7.10. I chose to use filled gray-colored bubbles. You may need to experiment with different values of the minimum and maximum bubble sizes, BRADIUSMIN and BRADIUSMAX, to obtain a visually pleasing plot that distinguishes points.

[1]The SURVEYREG procedure will also create a bubble plot in which the size of the bubble is proportional to the weight of each observation (see code in `nhanesgraphs.sas`), but does not have as many options for customizing the plot as the SGPLOT procedure.

Code 7.10. Draw the bubble plot (nhanesgraphs.sas).

```
PROC SGPLOT DATA = bubbleage;
  BUBBLE X = ridageyr Y = bmxbmi SIZE = sumwts / FILL FILLATTRS = (COLOR=gray)
    BRADIUSMIN = 1px BRADIUSMAX = 8px;
  XAXIS LABEL = "Age (years)" MIN = 0 MAX = 80;
  YAXIS LABEL = "Body Mass Index" MIN = 10 MAX = 70;
  TITLE 'Bubble plot, size proportional to sum of observation weights';
RUN;
```

Output 7.10. Bubble plot (nhanesgraphs.sas).

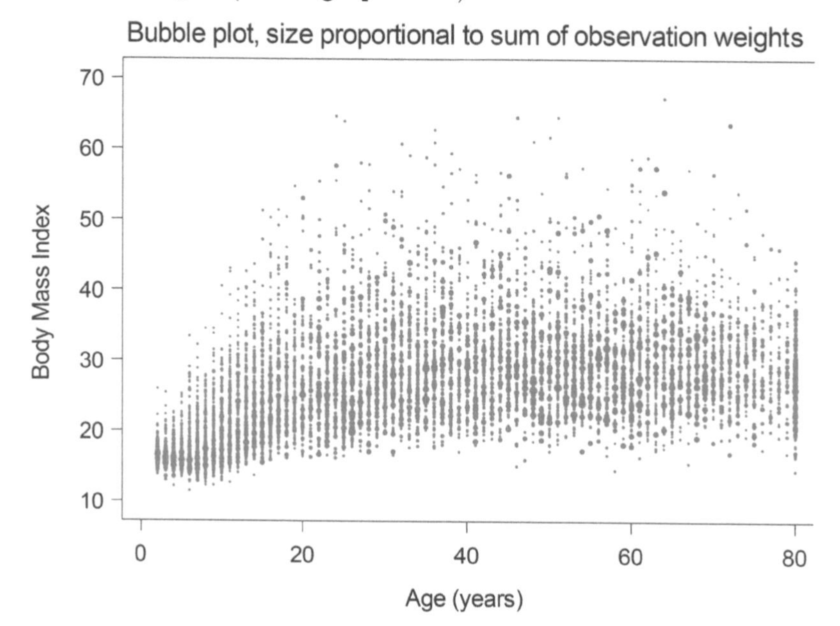

Binned bubble plot. A binned bubble plot is similar in spirit to the plot shown in Output 7.10, except now the weights are summed for observations in bins rather than just the distinct pairs (x, y). Code 7.11 defines variables *bmigroup* and *agegroup*, which round the values of *bmxbmi* and *ridageyr* to the nearest multiple of 5. Then, Code 7.12 draws the plot using the SGPLOT procedure.

Code 7.11. Binned bubble plot: compute bins (nhanesgraphs.sas).

```
DATA groupage;
   SET nhanes;
   bmigroup = ROUND(bmxbmi,5);
   agegroup = ROUND(ridageyr,5);

PROC SORT DATA = groupage;
   BY bmigroup agegroup;

/* Now sum the weights for each bin of BMI value and age value */
PROC MEANS DATA = groupage NOPRINT SUM;
   BY bmigroup agegroup;
   VAR  wtmec2yr;
   OUTPUT OUT = circleage SUM = sumwts;
```

Code 7.12. Binned bubble plot: draw the plot (**nhanesgraphs.sas**).

```
PROC SGPLOT DATA = circleage;
  BUBBLE X = agegroup Y = bmigroup SIZE = sumwts / FILL BRADIUSMAX = 16px
     BRADIUSMIN = 1px;
  XAXIS LABEL = "Age (years)" MIN = 0 MAX = 80;
  YAXIS LABEL = "Body Mass Index" MIN = 10 MAX = 70;
  TITLE 'Binned bubble plot, area proportional to sum of bin weights';
RUN;
```

Output 7.12. Binned bubble plot (**nhanesgraphs.sas**).

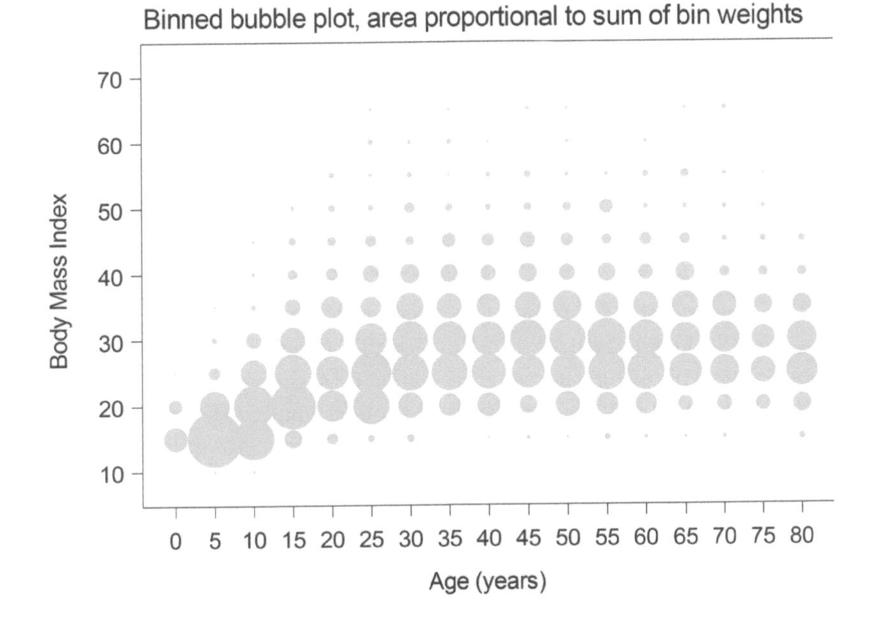

Plot subsample of data. Code 7.13 draws the scatterplot of a subsample that has been selected with the SURVEYSELECT procedure. The subsample is selected with probability proportional to the weights, so it is self-weighting. Here, METHOD=pps was used to select the subsample, but METHOD=poisson or METHOD=pps_wr from Table 6.1 could also be used.

Code 7.13. Plot probability-proportional-to-weights subsample (**nhanesgraphs.sas**).

```
PROC SURVEYSELECT DATA=nhanes OUT=nhanesplot SAMPSIZE=500 METHOD=pps SEED=38274;
  SIZE wtmec2yr;
PROC SGPLOT DATA = nhanesplot;
  SCATTER X = ridageyr Y = bmxbmi / MARKERATTRS = (SYMBOL=Plus COLOR=black);
  XAXIS LABEL = "Age (years)" MIN = 0 MAX = 80;
  YAXIS LABEL = "Body Mass Index" MIN = 10 MAX = 70;
```

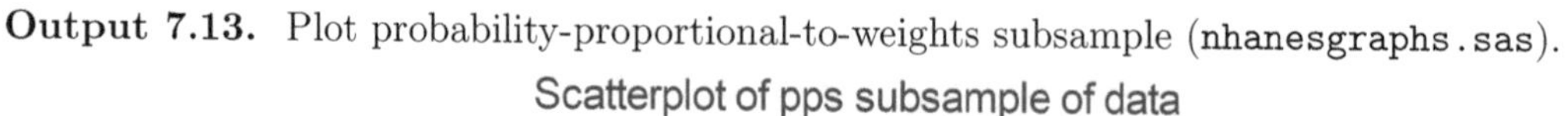

Output 7.13. Plot probability-proportional-to-weights subsample (`nhanesgraphs.sas`).

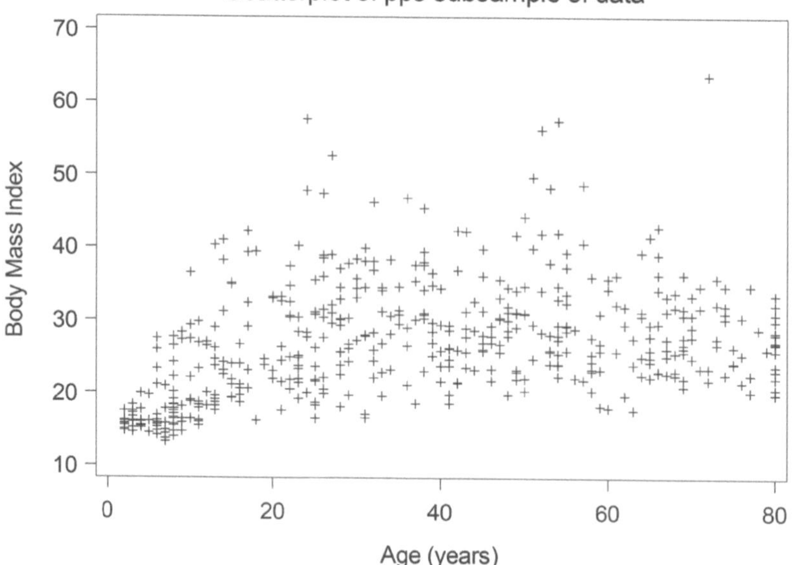

Shaded plot. There are several ways to create a shaded plot, sometimes called a heat map. The easiest uses the PLOTS option of the SURVEYREG procedure, which performs a regression of y on x incorporating the weights. Code 7.14 shows the commands used to obtain a plot where the shading of the hexagons is proportional to the sum of the weights. You can choose to plot rectangles if desired by typing SHAPE=rectangular. The graph also displays the regression line.

Code 7.14. Shaded plot from the SURVEYREG procedure (`nhanesgraphs.sas`).

```
PROC SURVEYREG DATA = nhanes PLOTS = (FIT(WEIGHT=heatmap SHAPE=hexagonal));
   WEIGHT wtmec2yr;
   STRATA sdmvstra;
   CLUSTER sdmvpsu;
   MODEL bmxbmi = ridageyr;
```

Output 7.14. Shaded plot from the SURVEYREG procedure (`nhanesgraphs.sas`).

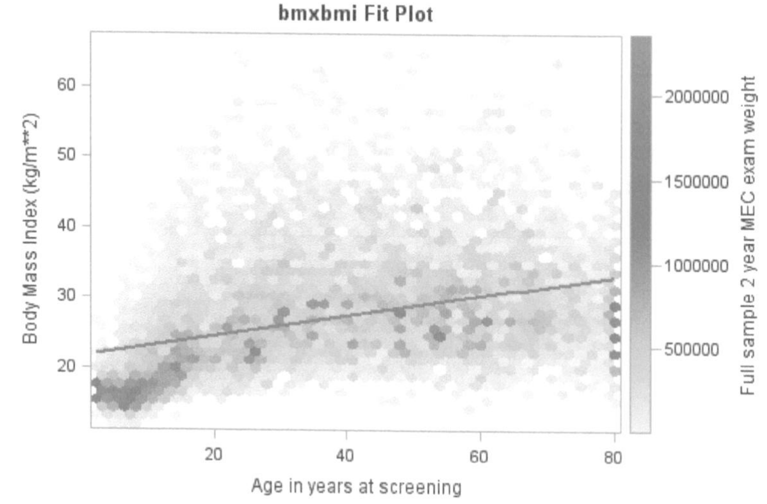

If you want more control over the plot, you can use the SGPLOT procedure. As with the binned bubble plot, you first need to calculate the sum of the weights in different bins. In Code 7.15, the values of *bmxbmi* are rounded to the nearest unit; the values of *ridageyr* are integers, so no further rounding is needed.

The HEATMAPPARM statement of the SGPLOT procedure draws the heat map. The intensity of shading is given by the COLORRESPONSE variable *sumwts*. I used a two-color scale because this book is printed in black-and-white, but other color schemes can be adopted.

Code 7.15. Shaded plot from the SGPLOT procedure (`nhanesgraphs.sas`).

```
DATA groupbmi;
   SET nhanes;
   BMIGROUP = ROUND(bmxbmi,1);
PROC SORT DATA = groupbmi;
   BY bmigroup ridageyr;
PROC MEANS DATA = groupbmi NOPRINT SUM;
   BY bmigroup ridageyr;
   VAR wtmec2yr;
   OUTPUT OUT = shadeplot SUM = sumwts;

PROC SGPLOT DATA = shadeplot NOAUTOLEGEND;
   HEATMAPPARM X = ridageyr Y = bmigroup COLORRESPONSE = sumwts / COLORMODEL =
    TwoColorRamp;
   XAXIS LABEL = "Age (years)" MIN = 0 MAX = 80;
   YAXIS LABEL = "Body Mass Index" MIN = 10 MAX = 70;
   TITLE  'Shaded plot of BMI vs age using SGPLOT procedure';
RUN;
```

Output 7.15. Shaded plot from the SGPLOT procedure (`nhanesgraphs.sas`).

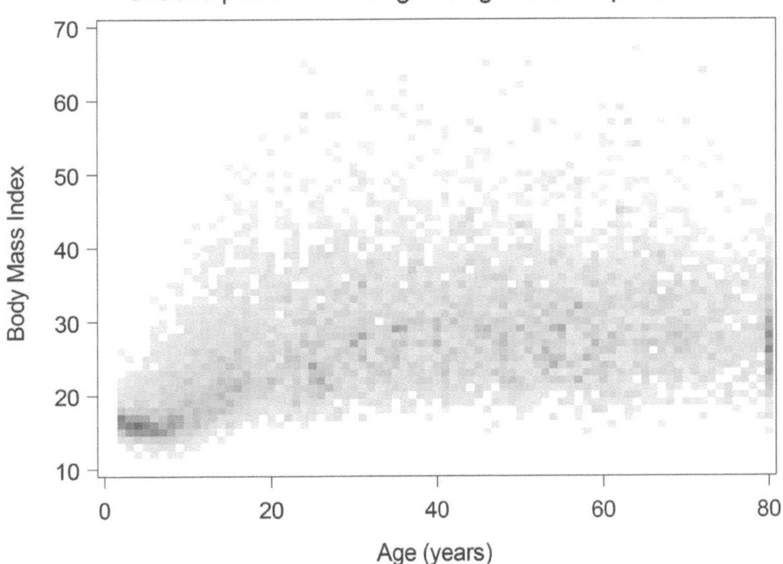

Side-by-side boxplots. The easiest way to obtain side-by-side boxplots for survey data is through the SURVEYMEANS procedure with the DOMAIN statement. Code 7.16

first defines the domains in the DATA step as the age groups (formed by rounding the variable *ridageyr* to the nearest 5), then requests the boxplots using PLOTS= in the SURVEYMEANS statement. Because the WEIGHT, STRATA, and CLUSTER statements are included, this code also produces the correct standard errors for all statistics.

The EXCLUDE option for the DOMAIN plot excludes the box displaying the distribution the full sample (without the EXCLUDE option, this is displayed to the left of the domain boxplots as in Output 7.6(b)).

Code 7.16. Side-by-side boxplots from the SURVEYMEANS procedure (`nhanesgraphs.sas`).

```
DATA groupage;
   SET nhanes;
   agegroup = round(ridageyr,5);

PROC SORT DATA = groupage;
  BY agegroup;

PROC SURVEYMEANS DATA = groupage PLOTS= (DOMAIN(EXCLUDE)) NOMCAR MEAN PERCENTILE=
     (0 25 50 75 100);
  WEIGHT wtmec2yr;
  STRATA sdmvstra;
  CLUSTER sdmvpsu;
  VAR bmxbmi;
  DOMAIN agegroup;
  TITLE 'Side-by-side boxplots';
RUN;
```

Output 7.16. Side-by-side boxplots from the SURVEYMEANS procedure (`nhanesgraphs.sas`).

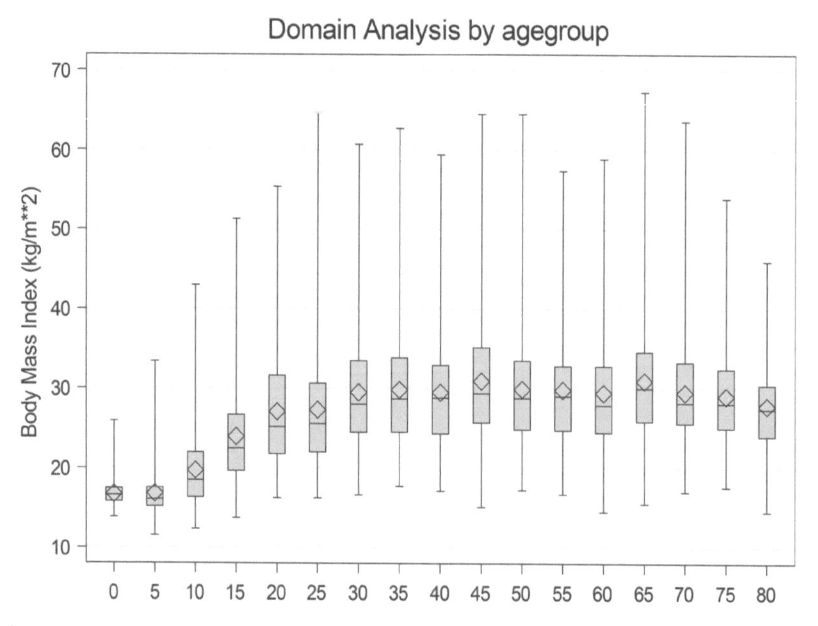

This boxplot can also be produced using the SGPLOT procedure, which has more options for customization. Create the variable *agegroup* and sort the data as in Code 7.16, then use the VBOX command in Code 7.17, including the WEIGHT option, to create the boxplots.

The option WHISKERPCT=0 extends the vertical lines to the minimum and maximum values of the data; this option is useful for large data sets.

Code 7.17. Side-by-side boxplots from the SGPLOT procedure (`nhanesgraphs.sas`).

```
PROC SGPLOT DATA = groupage;
  VBOX bmxbmi / WEIGHT = wtmec2yr CATEGORY = agegroup WHISKERPCT = 0;
  XAXIS LABEL = "Age (years)" MIN = 0 MAX = 80;
  YAXIS LABEL = "Body Mass Index" MIN = 10 MAX = 70;
  TITLE 'Side-by-side boxplots from the SGPLOT procedure';
RUN;
```

Output 7.17. Side-by-side boxplots from the SGPLOT procedure (`nhanesgraphs.sas`).

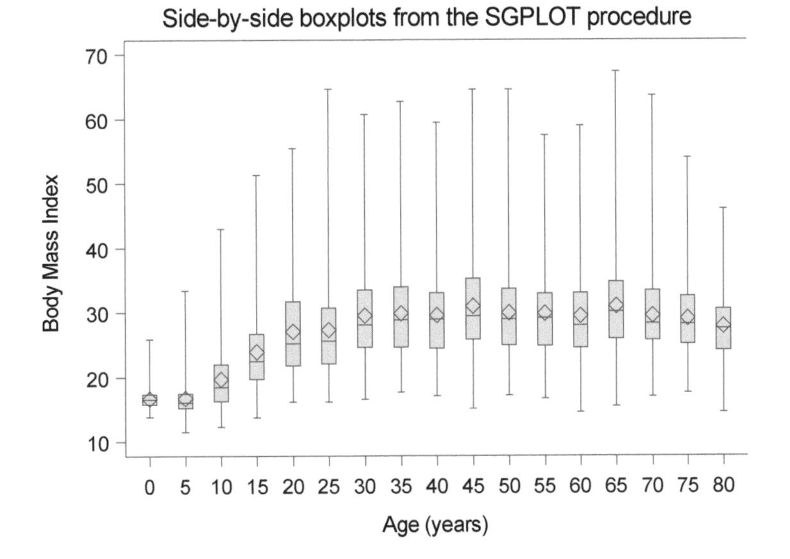

Smoothed trend line for mean. The SURVEYREG procedure will calculate smoothing splines. In essence, a cubic spline model specifies dividing the horizontal (x) axis into segments; the points where the segments join are called "knots." Then a cubic polynomial function is fit for each segment, with the constraint that the function and its first two derivatives are required to be continuous at the knot values (see Eilers and Marx, 2021).

Code 7.18. Calculate predicted values for smoothing spline (`nhanesgraphs.sas`).

```
PROC SURVEYREG DATA = nhanes PLOTS = NONE;
   WEIGHT wtmec2yr;
   EFFECT spl = SPLINE(ridageyr / BASIS=bspline KNOTMETHOD=percentiles(4));
   MODEL bmxbmi = spl;
   OUTPUT OUT = splinepred PRED = splpred;
```

The EFFECT statement in Code 7.18 requests that the procedure construct a cubic spline basis. The option KNOTMETHOD=percentiles(4) specifies that 4 knots are to be used, and these are to be placed at equally spaced percentiles. You can also choose the knots to be equally spaced along the x-axis or list your own values for the knots. The number of knots determines the smoothness of the curve. In general, the larger the data set, or the wigglier the function, the more knots you should use. You can experiment with the number of knots to obtain a function that looks like it has the right amount of smoothing, or use criteria described by McConville and Breidt (2013) and Zhang et al. (2015).

The regression model is then fit to the spline basis functions defined in the EFFECT statement (because this is done in the SURVEYREG procedure, the survey weights are used), and the predicted values from the spline model are output to variable *splpred* in data set *splinepred.*

Now the smoothed curve can be graphed, either by itself or superimposed on one of the other scatterplots. Code 7.19 sorts the data set of the spline predictions by the x variable (so the plotted line will be smooth) and then concatenates it with the data set created in Code 7.9 so that the trend line can be plotted along with the bubble plot of the data. The color of the bubbles is set as "Gainsboro" for better visibility of the trend line.

After the concatenation, the observations from data set *bubbleage* have missing values for variable *splpred*, so the SERIES statement draws the trend line using only the points from data set *splinepred.* Similarly, the observations from data set *splinepred* have missing values for the variable *sumwts*, so the BUBBLE statement draws bubbles only for the data from set *bubbleage.* If your Y variables from the two data sets have the same name, rename one of them so that the statements in the SGPLOT procedure plot the appropriate data.

Code 7.19. Smoothed trend line with bubble plot (`nhanesgraphs.sas`).

```
PROC SORT DATA = splinepred;
   BY ridageyr;
DATA splineplot;  /* Concatenate the data sets */
   SET bubbleage splinepred;

PROC SGPLOT DATA = splineplot NOAUTOLEGEND;
   BUBBLE X = ridageyr Y = bmxbmi SIZE = sumwts / FILL FILLATTRS=(COLOR=Gainsboro)
    BRADIUSMAX = 8px BRADIUSMIN = 1px;
   SERIES X = ridageyr Y = splpred;
   XAXIS LABEL = "Age (years)" MIN = 0 MAX = 80;
   YAXIS LABEL = "Body Mass Index" MIN = 10 MAX = 70;
   TITLE 'Bubble plot with smoothed trend line';
```

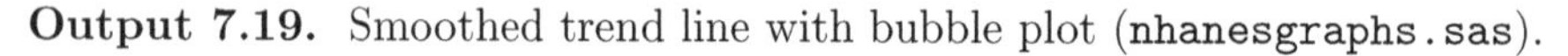

Output 7.19. Smoothed trend line with bubble plot (`nhanesgraphs.sas`).

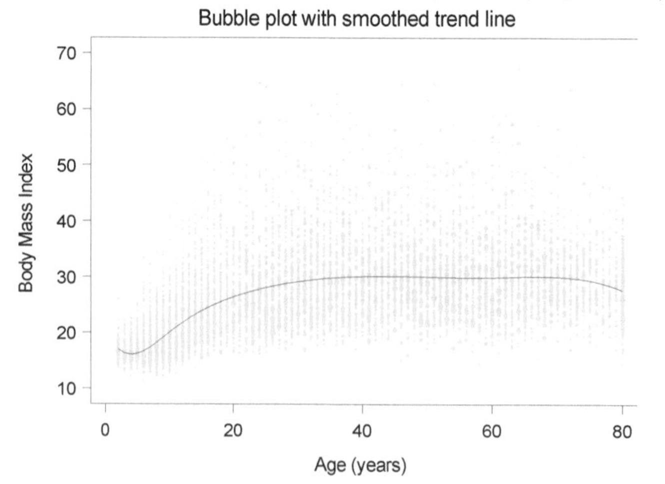

Smoothed trend lines for quantiles. The QUANTREG procedure fits quantile regression models to non-survey data. As in Code 7.18, the QUANTREG procedure can fit a smoothing spline to the quantiles by using the EFFECT statement. Instead of fitting smoothed lines to the means of different groups, however, the procedure fits separate lines estimating each conditional quantile (Koenker, 2005). You can think of this as "connecting the dots" of the quantiles shown in the side-by-side boxplots in Output 7.17.

The only difference between Code 7.20 and the code used to create the trend line going through the means is that the MODEL statement specifies which quantiles are desired.[2] I requested the quantiles corresponding to probabilities 0.05, 0.25, 0.5, 0.75, and 0.95, a range of quantiles that gives a good picture of the center (median) of the data, as well as the large and small values. The predicted values in output data set *qrplot* are named *qrpred1*, *qrpred2*, ..., *qrpred5*.

The SGPLOT procedure then draws the five lines. The SERIES statement connects the points in the order in which they appear in the data set, so be sure to first sort the data set by the x variable. The NOAUTOLEGEND option suppresses the legend on the plot.

Code 7.20. Smoothed quantile trend lines (`nhanesgraphs.sas`).

```
PROC QUANTREG DATA = nhanes  PLOTS=NONE;
   WEIGHT wtmec2yr;
   EFFECT spl = SPLINE(ridageyr / BASIS=bspline KNOTMETHOD=percentiles(4));
   MODEL bmxbmi = spl / QUANTILE = 0.05 0.25 0.5 0.75 0.95;
   OUTPUT OUT = qrplot PREDICTED = qrpred;
   TITLE 'Fit quantile regression with smoothing splines';

PROC SORT DATA=qrplot;
   BY ridageyr;

PROC SGPLOT DATA = qrplot NOAUTOLEGEND;
   SERIES X = ridageyr Y = qrpred1;
   SERIES X = ridageyr Y = qrpred2;
   SERIES X = ridageyr Y = qrpred3;
   SERIES X = ridageyr Y = qrpred4;
   SERIES X = ridageyr Y = qrpred5;
   XAXIS LABEL = "Age (years)" MIN = 0 MAX = 80;
   YAXIS LABEL = "Body Mass Index" MIN = 10 MAX = 70;
   TITLE 'Smoothed quantile trend lines';
RUN;
```

[2]Quantile regression is more complex computationally than the linear regression models fit by the SURVEYREG procedure. Sometimes the estimates do not converge at first, especially if there are only a few distinct values of y for some ranges of x. Always check the log when you use the QUANTREG procedure to make sure the estimates converged. If they do not converge, try adding the option `ALGORITHM = INTERIOR(TOLERANCE = 1e-4)` to the PROC QUANTREG statement so that a less "fussy" algorithm is used to do the computations.

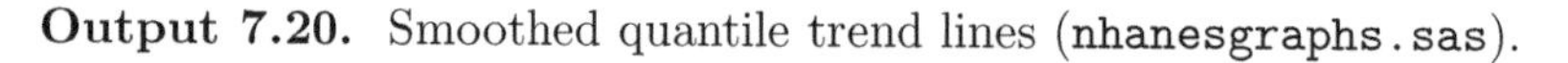

Output 7.20. Smoothed quantile trend lines (`nhanesgraphs.sas`).

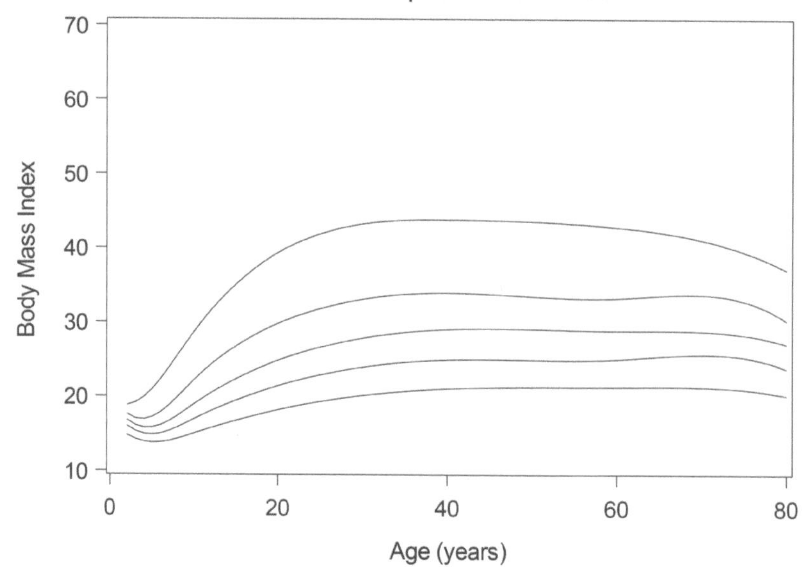

Customizing graphs. This chapter makes use of the graphics features of the SURVEYMEANS and SURVEYREG procedures available in version 15.1 of SAS software. For the other graphs, I developed code that calculates summary statistics and then draws the graph with the SGPLOT procedure. Future versions of the SURVEYMEANS and SURVEYREG may include some of these additional graphs; if that occurs, the additional steps I went through to create graphs such as the bubble plot with trend line may not be needed.

The graphs shown in this section use the default settings of the graphical procedures in SAS software. My goal was to show you how to create graphs for survey data without distraction from formatting issues. But SAS software offers numerous options for customizing graph appearance. You can change the fonts, axes, colors, titles, and almost anything else you would like. For more information, see the online documentation for the TEMPLATE procedure at `https://support.sas.com`, the article by Matange (2014), or the books by Matange and Heath (2011) and Matange and Bottitta (2016).

7.6 Additional Code for Exercises

Some of the exercises in Chapter 7 of SDA ask you to construct an empirical cumulative distribution function (ecdf) or an empirical probability mass function (epmf). Any population characteristic can be estimated using these functions. SAS software provides procedures that calculate commonly requested statistics such as means, totals, quantiles, and regression coefficients from survey data, so the survey analyst typically does not calculate or graph the ecdf. For continuous variables, it is usually more informative to create a histogram or a smoothed density estimate than to view the often-spiky epmf.

The ecdf and epmf are useful concepts for learning about how survey weights work, however, because they illustrate how the sample is used to create a reconstruction of the population. And they are easy to create in SAS software.

Example 7.5 of SDA. Code 7.21 calculates the epmf for the stratified sample of heights in `htstrat.csv`. I use the SURVEYMEANS procedure to do this because it is an old friend, but you could also use the MEANS or UNIVARIATE procedure. The variable *height* is declared to be categorical in the CLASS statement, so the statistics keyword MEAN requests the estimated population proportion for each of the distinct values of *height*; the estimated proportion for value y is the sum of the weights for observations having $height = y$ divided by the sum of all the weights in the sample. Since the empirical cdf and epmf calculations do not require standard errors, I could omit the STRATA statement from Code 7.21, but it does not hurt anything to leave it in. The code saves the values of $\hat{f}(y)$ in variable *mean* of data set *htstrat_epmf*.

Code 7.21. Calculate epmf for stratified sample of heights (`example0705.sas`).

```
PROC SURVEYMEANS DATA = htstrat MEAN;
   CLASS height;
   WEIGHT wt;
   STRATA gender;
   VAR height;
   ODS OUTPUT STATISTICS = htstrat_epmf;
RUN;
```

We then need to do a bit of data manipulation to get the statistics, in data set *htstrat_epmf*, into a form where they can be graphed. Code 7.22 does this. It renames the variable containing $\hat{f}(y)$ as *epmf*, and calculates *ecdf* as the cumulative sum of *epmf*. It also uses the INPUT function to convert the character variable *varlevel*, which was output from the SURVEYMEANS procedure to hold the levels of the CLASS variable, to a numeric variable *htval*.

Code 7.22. Create variables to be used in epmf plot (`example0705.sas`).

```
DATA htstrat_plot;
   retain ecdf 0;
   SET htstrat_epmf (RENAME = (mean = epmf));
   htval = INPUT(varlevel,best12.);
   zero = 0;
   ecdf = ecdf + epmf;
RUN;
```

The data in *htstrat_plot* are now in a form that can be used to draw the epmf. Code 7.23 uses the HIGHLOW statement, which draws vertical segments from the LOW value to the HIGH value for each value of X, to draw the lines.

Code 7.23. Graph of epmf for stratified sample of heights (`example0705.sas`).

```
PROC SGPLOT DATA = htstrat_plot;
   HIGHLOW X = htval HIGH = epmf LOW = zero;
   XAXIS LABEL = "Height value (cm)" MIN = 136 MAX = 206;
   YAXIS LABEL = "Empirical probability mass function";
   TITLE "Empirical pmf for stratified sample of heights";
RUN;
```

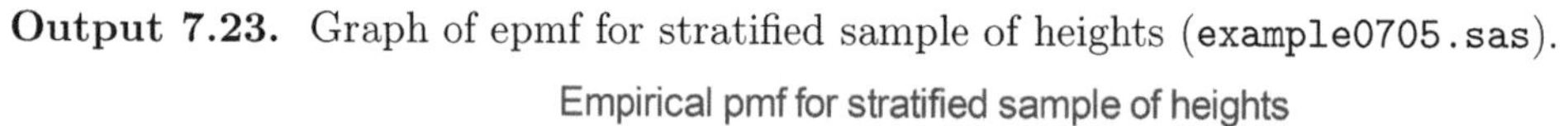

Output 7.23. Graph of epmf for stratified sample of heights (`example0705.sas`).

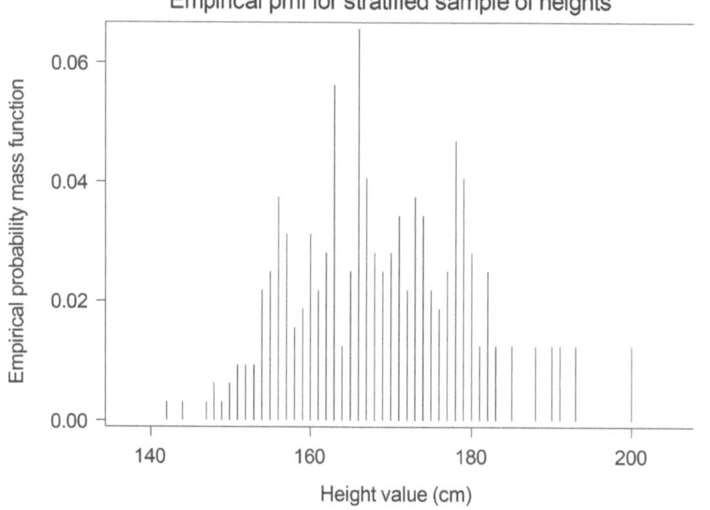

Code 7.24 draws an interpolated ecdf; the file `example0705.sas` also contains code for drawing the step function of the non-interpolated ecdf if that is desired.

Code 7.24. Graph of ecdf for stratified sample of heights (`example0705.sas`).

```
PROC SGPLOT DATA = htstrat_plot;
   SERIES x = htval y = ecdf;
   XAXIS LABEL = "Height value (cm)" MIN = 136 MAX = 206;
   YAXIS LABEL = "Empirical cdf";
   TITLE "Interpolated empirical cdf for stratified sample of heights";
RUN;
```

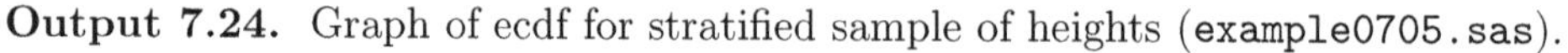

Output 7.24. Graph of ecdf for stratified sample of heights (`example0705.sas`).

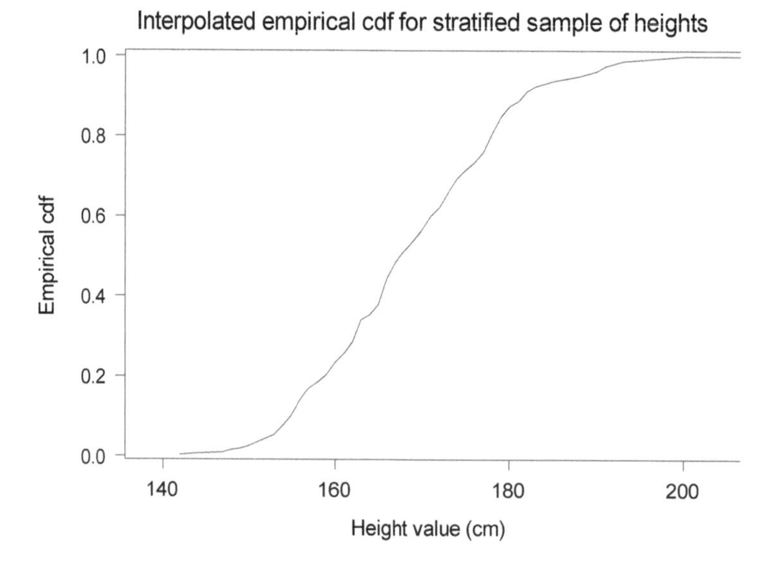

7.7 Summary, Tips, and Warnings

The SURVEYSELECT procedure can be used to select stratified multistage samples in which the psus are selected with either equal or unequal probabilities. The selection procedure takes place in stages, where a stratified sample of psus is drawn at the first stage. The code shown below contains the commands needed to select the first-stage sample.

```
PROC SURVEYSELECT DATA = population_data OUT = sample_data SAMPSIZE = sample_size
    METHOD = pps_method STATS SEED = seed_value;
   STRATA stratification_variable_for_psus;
   SIZE size_variable;
RUN;
```

The SURVEYMEANS, SURVEYREG, and SURVEYFREQ procedures all calculate statistics for stratified multistage samples. They, and the other SAS software procedures for calculating estimates from survey data, all use the WEIGHT, STRATA, and CLUSTER statements to specify the survey design. This is shown below for the SURVEYMEANS procedure.

```
PROC SURVEYMEANS DATA=sample_dataset PLOTS=ALL statistics_keywords;
   CLASS class_var1 ... class_varp;
   WEIGHT weight_variable;             /* Always include a weight variable */
   STRATA stratification_variables; /* Describes the stratification of the psus */
   CLUSTER cluster_variable;           /* The cluster_variable identifies the psus */
   VAR var1 var2 ... vark class_var1 ... class_varp;
   /* var1 var2 ... vark are the numeric variables to be analyzed */
   /* class_var1 ... class_varp are the categorical variables to be analyzed */
```

In general, I recommend omitting the TOTAL= option, which calculates a finite population correction. The survey procedures with the TOTAL= option give the correct variances for without-replacement sampling when one-stage cluster sampling is used and psus are selected with equal probabilities within strata. Otherwise, these variances are approximations that may be underestimates.

Quantiles are calculated in the SURVEYMEANS statement through the QUANTILE or PERCENTILE keywords.

The CLASS statement in the SURVEYMEANS procedure is optional; include it if you are analyzing any categorical variables. Additional optional statements for the SURVEYMEANS procedure include:

```
   DOMAIN domain_variables;
   RATIO 'label' numerator_var / denominator_var;
   POSTSTRATA poststratification_variables;
```

The SURVEYFREQ procedure has similar syntax, except that the CLASS statement is not used (all variables analyzed by the procedure are assumed to be categorical), and there is no DOMAIN statement.

```
PROC SURVEYFREQ DATA=sample_dataset;
   WEIGHT weight_variable;          /* Always include a weight variable */
   STRATA stratification_variables; /* Describes the stratification of the psus */
   CLUSTER cluster_variable;        /* The cluster_variable identifies the psus */
   TABLES class_var1 ... class_varp / CL CLWT DEFF;
   TABLES domain_var*class_var / CL ROW(DEFF);
```

The first TABLES statement requests percentages in one-way frequency tables of categorical variables *class_var1 ... class_varp*, along with design effects (DEFF) and confidence intervals for the percentages (CL) and estimated totals (CLWT). Clopper-Pearson intervals will be produced if you request option CL(TYPE=clopperpearson).

The second TABLES statement requests separate percentages and estimated totals of variable *class_var* for each level of the domain variable *domain_var*. Chapter 10 will discuss the use of the SURVEYFREQ procedure for producing and analyzing multiway tables.

The SURVEYREG procedure for stratified multistage samples will be discussed in Chapter 11.

The SURVEYMEANS and SURVEYREG procedures produce graphs of survey data that account for the weights. Additional graphs can be produced by computing summary statistics at individual points using the weights, then using the SGPLOT procedure.

This section lists only some of the statements and options available for the SURVEYMEANS, SURVEYFREQ, and SURVEYSELECT procedure. The procedures have many other capabilities and options, and these are described in SAS Institute Inc. (2021).

Tips and Warnings

- Always include WEIGHT, STRATA, and CLUSTER statements to analyze data from a stratified multistage sample. Check that the sum of the weights output by the procedure approximately equals the population size. Complex surveys often have many steps of weight calculations, and it is easy to get something wrong when constructing them or to misunderstand a weight variable calculated by someone else.
- The STRATA statement for a stratified multistage sample describes the stratification of the psus, and the CLUSTER statement specifies the variable that defines psu membership. No information about the second and subsequent stages of sampling is needed.
- If you want separate statistics for domains in the SURVEYMEANS or SURVEYREG procedures, use the DOMAIN statement. Performing separate analyses of domains, either through use of the BY statement or through performing an analysis of a subset of the data, can lead to incorrect standard error calculations. To obtain separate percentages for domains in the SURVEYFREQ procedure, construct a cross-classified table with the domain categories as the row variable.
- If the data set has missing data, use the NOMCAR option to treat the records with data as a domain so that standard errors are estimated correctly.
- When graphing survey data, incorporate the survey weights into the graph if it is desired to display relationships in the population.

8

Nonresponse

We have already seen most of the features needed to select survey samples and to compute estimates of means, totals, and quantiles using the sampling weights and survey design.

Nonresponse affects almost all surveys, however. Even samples of inanimate objects such as audit records may have missing accounts or missing items for accounts. This chapter looks at how SAS software deals with missing data, and then discusses some of the options in the software for performing weighting adjustments and imputing missing data. It also points you to resources giving step-by-step guidance for dealing with nonresponse.

8.1 How the Survey Analysis Procedures Treat Missing Data

The survey analysis procedures in SAS software have uniform defaults for handling missing data. Observations with missing values for the WEIGHT variable are excluded from all analyses. If you include replicate weights through the REPWEIGHTS statement (see Chapter 9), observations with missing values for any of the replicate weight variables are also excluded.

Observations with missing values for variables listed in STRATA, CLUSTER, VAR, RATIO, POSTSTRATA, DOMAIN, TABLES, and MODEL statements are excluded from the analysis under the default settings for the survey analysis procedures. But the MISSING option in the PROC SURVEYMEANS (or PROC SURVEYFREQ, PROC SURVEYREG, or other survey analysis procedure) statement provides a way to include observations that are missing values for categorical variables in the analysis. The MISSING option directs the procedure to treat missing values in categorical variables as a separate category.

The data set `impute.csv` contains the data from Table 8.4 of SDA; it codes the missing values by −99. You must convert missing value codes to periods (for numeric variables) or blanks (for character variables) before analyzing the data so that SAS software recognizes the values as missing. Otherwise, the procedures will treat these observations as though $y = -99$ and include them in calculations; for this data set, that would yield negative estimated percentages of persons victimized by crime. In the SAS data set `impute.sas7bdat`, the missing values have been converted to periods.

Let's first see what happens when we analyze these data in the SURVEYMEANS procedure without the MISSING option. The variables *crime* and *violcrime* both take values 0 and 1, where 1 indicates victimization by crime or by violent crime, respectively. Proportions can be estimated for a binary variable in two ways: by treating the binary variable as numeric and finding the mean, or by treating it as a CLASS variable and calculating the proportion in each category. To illustrate the different treatment for numeric and categorical variables, Code 8.1 treats *crime*, with missing values for two observations, as numeric and it declares *violcrime*, with missing values for four observations, as a CLASS variable.

DOI: 10.1201/9781003160366-8

Code 8.1. Analyze data with missing values, no MISSING option (`example0810.sas`).

```
DATA impute;
   SET datalib.impute;
   WT = 100;

PROC SURVEYMEANS DATA = impute NOBS NMISS MEAN;
   CLASS violcrime;
   WEIGHT wt;
   VAR crime violcrime;
RUN;
```

Output 8.1. Analyze data with missing values, no MISSING option (`example0810.sas`).

Statistics

Variable	Level	N	N Miss	Mean	Std Error of Mean
crime		18	2	0.333333	0.114332
violcrime	0	13	0	0.812500	0.100778
	1	3	0	0.187500	0.100778

The data set has 20 observations. In Output 8.1, the estimated proportion for numeric variable *crime* is calculated from the 18 observations having non-missing values. The estimated proportions in each category for CLASS variable *violcrime* are calculated using the 16 observations having non-missing values.

Now let's see what happens with the MISSING option in Code 8.2. The output for numeric variable *crime* is identical. But three proportions are now estimated for CLASS variable *violcrime*, where the first category consists of the four observations with missing values (.) for the variable.

Code 8.2. Analyze data with missing values, with MISSING option (`example0810.sas`).

```
PROC SURVEYMEANS DATA = impute MISSING NOBS NMISS MEAN;
   CLASS violcrime;
   WEIGHT wt;
   VAR crime violcrime;
```

Output 8.2. Analyze data with missing values, with MISSING option (`example0810.sas`).

Statistics

Variable	Level	N	N Miss	Mean	Std Error of Mean
crime		18	2	0.333333	0.114332
violcrime	.	4	0	0.200000	0.091766
	0	13	0	0.650000	0.109424
	1	3	0	0.150000	0.081918

The survey analysis procedures also have a NOMCAR option that tells them to treat missing values as a domain for variance estimation purposes. This option was used in Code 7.4.

8.2 Poststratification and Weighting Class Adjustments

Unit nonresponse occurs when a unit selected for the sample provides no data. In some data sets, the nonrespondent unit may be represented in the data set but have missing values for all survey responses. Other data sets contain no records for nonrespondent units.

The most common method for trying to compensate for potential effects of nonresponse is to weight the data. The sampling frame may contain information that can be used in weighting class adjustments, or information known about the population from an external source may be used to poststratify or rake the data. In most large surveys, both weighting class adjustments and poststratification are used, as described in Section 8.6 of SDA.

The SURVEYMEANS procedure will construct poststratified weights to calibrate the survey to independent population estimates for the poststrata. Lewis (2012) provides a useful guide to using SAS software procedures to calculate nonresponse-adjusted weights, with algorithms for producing weighting-class adjustments, poststratified weights, and weights calculated through a logistic regression propensity model.

As we saw in Section 4.4, the POSTSTRATA statement in the SURVEYMEANS procedure creates poststratified weights, and the standard errors calculated by the SURVEYMEANS procedure creating those weights account for the poststratification. The poststratified weights output by the procedure, however, do not have enough information by themselves to allow calculation of the poststratified standard errors. Thus, if you subsequently use the poststratified weights produced by the SURVEYMEANS procedures in a regression analysis, the estimated regression coefficients produced by the SURVEYREG procedure will be calculated with the poststratified weights that you supplied, but the standard errors will not incorporate the effect of the poststratification weight adjustments. If you want to obtain correct standard errors for subsequent analyses using poststratified weights calculated by the SURVEYMEANS procedure, you should create replicate weights for variance estimation, as will be discussed in Chapter 9.

Macros for weight adjustments. The POSTSTRATA statement can create poststratified weights, and you can create replicate weights that allow variance estimates of characteristics calculated using the poststratified weights to account for the weight adjustments. However, many surveys have several stages of weighting class adjustments, poststratification, raking, or calibration. In general, the easiest way to incorporate the effects of all of those stages of nonresponse adjustments is to create a set of replicate weights for estimating variances, as described in Section 9.3.4.

Numerous authors have written SAS macros that perform other types of weight adjustments, and this section lists a few of the published macros available as of this writing in 2020.

The SAS macro `%SurveyCalibrate`, developed by An (2020), performs general calibration of survey weights. The user supplies the original weight variable, the population totals (known from an auxiliary data source) for the calibration variables, and information about the sampling design. The macro produces replicate weights that incorporate the effects of the calibration into the variance estimation.

The macro performs poststratification and more general types of calibration. If $\{x_{i1}, \ldots, x_{ip}\}$ is a set of variables measured for each unit i in the sample, the macro creates calibrated weights $w_i^* = g_i w_i$ that satisfy

$$\sum_{i \in \mathcal{S}} w_i^* x_{ij} = t_{xj}, \tag{8.1}$$

where t_{xj} is the population total for auxiliary variable j that is known from an external source. Poststratification is a special case of calibration where the variables x_{ij} indicate the poststrata. The macro allows several distance functions to be used for calculating the weights, and it also allows the user to specify minimum and maximum allowances for the adjustments g_i—for example, one might want to set a lower bound of 0.5 to avoid down-weighting observations too much (this also avoids the awkward problem of negative calibration weights) and an upper bound of 2 to avoid extreme weight adjustments. If the upper and lower bounds are set, though, the resulting weights will only approximately meet the calibration constraints in (8.1).

Although the SURVEYMEANS procedure does not, as of this writing, allow directly for raking, Izrael et al. (2000) developed a macro that calculates raked weights. The most recent version of the macro (Izrael et al., 2017) is available for free download from `https://www.abtassociates.com/raking-survey-data-aka-sample-balancing`. The macro by Rizzo (2014) calculates raked weights and trims excessively large weights.

Many national statistics agencies and private companies have developed customized software that performs multiple steps of weighting and calibration for surveys.

8.3 Imputation

Many data sets have item nonresponse in addition to unit nonresponse. Item nonresponse occurs when a unit has responses for some of the items on the survey but missing values for other items. A person may answer questions about age, race, gender, and health outcomes but decline to answer questions about income. Or the data editing process may uncover logical inconsistencies in the data, such as a 6-year-old who is listed as being married. If the discrepancy cannot be resolved, both age and marital status may be recoded as missing for that person.

Imputation fills in values for these missing items. It can also be used for unit nonresponse, where a unit chosen for the sample provides no data at all, but in this section I assume that weighting will be used for unit nonresponse and the imputation is limited to missing items.

Weighting-class adjustment methods require that the variables used in the weighting are known for all sample members, both respondent and nonrespondent. Poststratification requires the poststratification variables to be known for all respondents. But some respondents may have item nonresponse for some of the variables used in the weighting. For example, a respondent may answer some of the questions but refuse to give his age. In general, imputation must be used for variables used in weight calculations; it may be desired for other variables as well. SDA briefly mentions different methods that are available for imputation; for a more detailed description of methods, see Haziza (2009).

The SURVEYIMPUTE procedure, described in detail by Mukhopadhyay (2016), will perform several types of imputation. The next example illustrates the SURVEYIMPUTE procedure for a simple hot-deck imputation, in which a donor is selected randomly from a group of respondents who are similar on the non-missing items. The procedure also provides options for fractional imputation (Kim and Fuller, 2004; Yang and Kim, 2016), which is often the preferred method because it allows the survey analyst to account for the extra variability induced by imputation.

Example 8.10 of SDA. Code 8.3 uses the SURVEYIMPUTE procedure to perform a random hot-deck imputation for variables *crime* and *violcrime* in data *impute*. The CELLS statement defines the cells used for the imputation; here, the cross-classification of *gender* and *ageclass*. Note that this simple example does not involve stratification or clustering. For a stratified multistage sample, you should include STRATA and CLUSTER statements so that the SURVEYIMPUTE procedure accounts for the survey design.

Code 8.3. Perform random hot deck imputation in cells (`example0810.sas`).

```
DATA impute;
   SET impute;
   IF age <= 34 THEN ageclass = 1;
   ELSE IF age > 34 THEN ageclass = 2;

PROC SURVEYIMPUTE DATA = impute METHOD = hotdeck (SELECTION = srswr) SEED = 3817;
   CELLS gender ageclass;
   VAR crime violcrime;
   OUTPUT OUT = hotdeck DONORID = donor;
RUN;
```

Output 8.3(a). Missing data patterns and imputation summary (`example0810.sas`).

Imputation Information	
Data Set	WORK.IMPUTE
Imputation Method	HOTDECK
Selection Method	SRSWR
Random Number Seed	3817

Number of Observations Read	20
Number of Observations Used	20

Missing Data Patterns							Group Means	
Group	crime	violcrime	Freq	Sum of Weights	Unweighted Percent	Weighted Percent	crime	violcrime
1	X	X	15	15	75.00	75.00	0.266667	0.200000
2	X	.	3	3	15.00	15.00	0.666667	.
3	.	X	1	1	5.00	5.00	.	0
4	.	.	1	1	5.00	5.00	.	.

Imputation Summary		
Observation Status	Number of Observations	Sum of Weights
Nonmissing	15	15
Missing	5	5
Missing, Imputed	5	5
Missing, Not Imputed	0	0

Output 8.3(b). Results of hot deck imputation (`example0810.sas`).

Obs	UnitID	ImpIndex	donor	person	age	gender	education	crime	violcrime	wt	ageclass
1	1	0	.	1	47	M	16	0	0	100	2
2	2	0	.	2	45	F	.	1	1	100	2
3	3	0	.	3	19	M	11	0	0	100	1
4	4	0	.	4	21	F	.	1	1	100	1
5	5	0	.	5	24	M	12	1	1	100	1
6	6	0	.	6	41	F	.	0	0	100	2
7	7	1	16	7	36	M	20	1	0	100	2
8	8	0	.	8	50	M	12	0	0	100	2
9	9	1	18	9	53	F	13	0	0	100	2
10	10	1	14	10	17	M	10	1	0	100	1
11	11	0	.	11	53	F	12	0	0	100	2
12	12	0	.	12	21	F	12	0	0	100	1
13	13	1	20	13	18	F	11	1	0	100	1
14	14	0	.	14	34	M	16	1	0	100	1
15	15	0	.	15	44	M	14	0	0	100	2
16	16	0	.	16	45	M	11	0	0	100	2
17	17	0	.	17	54	F	14	0	0	100	2
18	18	0	.	18	55	F	10	0	0	100	2
19	19	1	20	19	29	F	12	0	0	100	1
20	20	0	.	20	32	F	10	0	0	100	1

Multiple imputation in SAS software. The MI procedure (MI stands for "multiple imputation") performs other types of imputations. Berglund and Heeringa (2014) describe how to use the procedure with complex survey data. Berglund (2015) illustrates the MI procedure with data from the National Comorbidity Survey–Replication, a stratified multistage sample about mental health in the United States.

8.4 Summary, Tips, and Warnings

SAS software provides multiple methods for dealing with missing data. The POSTSTRATA statement in the SURVEYMEANS procedure can be used to created nonresponse-adjusted weights. The SURVEYIMPUTE procedure imputes missing values in survey data using various hot-deck procedures. The MI procedure can be used to perform multiple imputation with survey data. In addition, many macros have been published that perform specific weighting tasks such as raking or general calibration.

Tips and Warnings

- Check how missing values are coded before analyzing your data set, and recode the missing values to those used by SAS software.
- When creating poststratified weights in the SURVEYMEANS procedure that will be used for other analyses, also create replicate weights that can be used for variance estimation. These will be discussed in Chapter 9.

9

Variance Estimation in Complex Surveys

SAS software provides multiple ways of estimating variances from complex surveys. Linearization methods may be used if stratification and clustering variables are available. Some large surveys come with columns of replicate weights, and the survey procedures in SAS software will calculate variance estimates for standard replication methods as well. The survey analysis procedures also allow you to construct your own set of replicate weights for a survey from the stratification and clustering information.

9.1 Linearization (Taylor Series) Methods

By default, the survey analysis procedures in SAS software use linearization methods to calculate variances. You can request this explicitly by specifying VARMETHOD=TAYLOR in the main SURVEYMEANS (or any other survey analysis procedure) statement. Code 9.1, which explicitly requests linearization variance estimates, gives the same output as Code 3.11.

Code 9.1. Explicitly request linearization variance estimates for the example in Code 3.11.

```
PROC SURVEYMEANS DATA = agstrat TOTAL=strattot VARMETHOD = TAYLOR PLOTS=ALL MEAN
     SUM CLM CLSUM DF;
   WEIGHT strwt;
   STRATA region;
   VAR acres92;
RUN;
```

9.2 Replicate Samples and Random Groups

The random group methods in Section 9.2 of SDA are seldom used in practice. But they can be of interest because they motivate how replication methods work in general, and are sometimes useful for providing quick-and-easy variance estimates to check calculations for a complex survey. The methods require three steps:

1. Select the replicate samples from the population, or divide the probability sample among the random groups.
2. Calculate the statistic of interest from each replicate, using the survey weights.
3. Use the estimated statistics from the replicates to calculate the standard error.

DOI: 10.1201/9781003160366-9

Example 9.3 of SDA. The SURVEYSELECT procedure allows you to carry out the first step and select replicate samples from a population. Code 9.2 defines data set *public_college* to consist of the 500 public colleges and universities. It then selects REPS=5 independent simple random samples (SRSs), each of size 10, from the data set and places the samples in data set *collegerg*. The variable *repgroup* is designated by the REPNAME option to contain the replicate number.

Code 9.2. Select 5 replicate samples (`example0903.sas`).

```
DATA public_college;
   SET datalib.college;
   IF control = 2 THEN delete; /* Remove non-public schools from the data set */

PROC SURVEYSELECT DATA = public_college METHOD=srs SAMPSIZE=10 REPS=5 (REPNAME =
    repgroup) SEED=38572 STATS OUT=collegerg;

PROC PRINT DATA = collegerg;
   WHERE repgroup = 1;
   VAR repgroup samplingweight instnm city stabbr tuitionfee_in tuitionfee_out;
RUN;
```

Output 9.2(a). Select 5 replicate samples (`example0903.sas`).

The SURVEYSELECT Procedure

Selection Method	Simple Random Sampling

Input Data Set	PUBLIC_COLLEGE
Random Number Seed	38572
Sample Size	10
Selection Probability	0.02
Sampling Weight	50
Number of Replicates	5
Total Sample Size	50
Output Data Set	COLLEGERG

Output 9.2(a), created by the SURVEYSELECT procedure, confirms that 5 replicate SRSs, each of size 10, were selected from the population. An observation in replicate sample k has sampling weight 50 ($= 500/10$), and this weight is stored in the output data file *collegerg* in variable *SamplingWeight*.

The PRINT procedure shows the values of the selected variables for the colleges in replicate sample 1, in Output 9.2(b). Note that the weights for each replicate sample sum to the population size, 500.

Output 9.2(b). Print replicate sample 1 (`example0903.sas`).

Obs	repgroup	SamplingWeight	INSTNM	CITY	STABBR	TUITIONFEE_IN	TUITIONFEE_OUT
1	1	50	University of Louisiana at Lafayette	Lafayette	LA	9912	23640
2	1	50	CUNY City College	New York	NY	7140	14810
3	1	50	Binghamton University	Vestal	NY	9808	26648
4	1	50	University of North Carolina at Chapel Hill	Chapel Hill	NC	8987	35170
5	1	50	Oklahoma Panhandle State University	Goodwell	OK	7930	8674
6	1	50	University of Science and Arts of Oklahoma	Chickasha	OK	7200	17550
7	1	50	Rhode Island College	Providence	RI	8929	21692
8	1	50	Francis Marion University	Florence	SC	11976	22488
9	1	50	East Tennessee State University	Johnson City	TN	8935	27199
10	1	50	Stephen F Austin State University	Nacogdoches	TX	8316	18276

For Step 2, compute the statistic of interest from each replicate sample, as in Code 9.3. First, sort the data set by *repgroup*. Then, calculate the ratio of average out-of-state tuition to average in-state tuition for each replicate sample by running the SURVEYMEANS procedure with the BY statement. This performs the analysis separately for each replicate (output not shown). The ratios calculated from the 5 analyses are exported to the data set called *Ratiostats*.

The survey weights are the only design feature used for these calculations. Even if a replicate sample has stratification or clustering, that information is not needed to calculate the point estimate of the parameter of interest θ for the replicates. The effect of the stratification or clustering on the variance is incorporated in the variability among the replicate estimates $\hat{\theta}_r$. For example, if clustering decreases the precision for $\hat{\theta}_r$, then the estimates $\hat{\theta}_r$ will vary more from replicate to replicate, and the decreased precision will show up in a large value of the sample variance for the replicate values $\hat{\theta}_r$.

Code 9.3. Analyze replicate samples (`example0903.sas`).

```
PROC SORT DATA=collegerep;
   by repgroup;

PROC SURVEYMEANS DATA=collegerep PLOTS=NONE MEAN;
   BY repgroup;
   WEIGHT samplingweight;
   VAR tuitionfee_in tuitionfee_out;
   RATIO 'out-of-state to in-state' tuitionfee_out / tuitionfee_in;
   ODS OUTPUT STATISTICS = Outstats RATIO = Ratiostats;

PROC PRINT DATA=Ratiostats;
```

Output 9.3 shows the values of $\hat{\theta}_r$ for the 5 replicate samples in the *Ratio* column. These values are the only part of the output used in Step 3; ignore the values given in the *StdErr* column.

Output 9.3. Analyze replicate samples; printout of data set *ratiostats* (`example0903.sas`).

Obs	repgroup	RatioLabel	NumeratorName	NumeratorLabel	DenominatorName	DenominatorLabel	Ratio	StdErr
1	1	out-of-state to in-state	TUITIONFEE_OUT	TUITIONFEE_OUT	TUITIONFEE_IN	TUITIONFEE_IN	2.424994	0.231178
2	2	out-of-state to in-state	TUITIONFEE_OUT	TUITIONFEE_OUT	TUITIONFEE_IN	TUITIONFEE_IN	2.252934	0.236582
3	3	out-of-state to in-state	TUITIONFEE_OUT	TUITIONFEE_OUT	TUITIONFEE_IN	TUITIONFEE_IN	2.088359	0.108107
4	4	out-of-state to in-state	TUITIONFEE_OUT	TUITIONFEE_OUT	TUITIONFEE_IN	TUITIONFEE_IN	2.051040	0.219564
5	5	out-of-state to in-state	TUITIONFEE_OUT	TUITIONFEE_OUT	TUITIONFEE_IN	TUITIONFEE_IN	2.412988	0.244254

For Step 3, treat the 5 estimated ratios in Output 9.3 as independent and identically distributed observations, and calculate their mean $\tilde{\theta}$ and standard error. Code 9.4 calculates the 95% confidence interval in Output 9.4 as $\tilde{\theta} \pm t\,\text{SE}(\tilde{\theta})$, where t is the critical value from a t distribution with 4 (number of replicates minus one) degrees of freedom (df). I used the SURVEYMEANS procedure to do the calculations, but the MEANS or TTEST procedure could also be used.

Code 9.4. Calculate the mean and variance of the set of 5 ratios calculated from the replicate samples (`example0903.sas`).

```
PROC SURVEYMEANS DATA=Ratiostats PLOTS=NONE MEAN VAR CLM DF;
   VAR Ratio;
   TITLE 'Calculate the mean of the ratios from the 5 replicate samples, with CI';
RUN;
```

Output 9.4. Calculate the mean and variance of the set of 5 ratios calculated from the replicate samples (`example0903.sas`).

The SURVEYMEANS Procedure

Data Summary	
Number of Observations	5

Statistics						
Variable	DF	Mean	Std Error of Mean	Var of Mean	95% CL for Mean	
Ratio	4	2.246063	0.078368	0.006142	2.02847769	2.46364864

I illustrated the method for estimating a ratio, but the same method can be used for any statistic you would like to estimate. The method can also be applied to statistics that are not calculated by the survey analysis procedures. All you need to do is to calculate the statistic of interest for each replicate using the survey weights, then apply t confidence interval methods to the statistics calculated from the replicates.

Random groups. The procedure is exactly the same for random groups created from an existing survey: Calculate the value of the statistic of interest for each random group, then calculate the mean and standard error of the replicate statistics. If you are estimating population totals, scale the weights for each random group so they sum to the population size.

Example 9.4 of SDA. The variable *randgrp* defines the random groups for the Survey of Youth in Custody data in `syc.csv`. The BY statement in Code 9.5 tells the SURVEYMEANS procedure to compute the mean and quartiles separately for each random group. Each is calculated using the WEIGHT in variable *finalwt*; we do not need to specify the stratification and clustering information because we do not use the standard errors produced by the SURVEYMEANS procedure. The statistics are output to data set *sycrgp*; only the point estimates, printed in the PRINT statement, are needed to calculate the mean of the replicate values $\tilde{\theta}$ and the confidence interval requested by the MEANS statement.

Code 9.5. Calculate variances using the random group method (`example0904.sas`).

```
PROC SURVEYMEANS DATA = syc MEAN PERCENTILE = (25 50 75);
  BY randgrp;
  WEIGHT finalwt;
  VAR age;
  ODS OUTPUT STATISTICS = sycrgp;

PROC PRINT DATA = sycrgp;
   VAR randgrp Mean Pctl_25 Pctl_50 Pctl_75;

PROC MEANS DATA = sycrgp MEAN VAR STD CLM;
   VAR Mean Pctl_50;
RUN;
```

Output 9.5(a). Print the statistics calculated from each random group (`example0904.sas`).

Obs	randgrp	Mean	Pctl_25	Pctl_50	Pctl_75
1	1	16.549468	14.7662	15.7698	17.0595
2	2	16.663307	15.0728	16.0247	16.9872
3	3	16.825441	14.9278	15.9102	17.4120
4	4	16.056877	14.1550	15.2881	16.7987
5	5	16.317757	14.2084	15.6707	16.9305
6	6	17.027976	15.1538	16.2252	17.7634
7	7	17.266045	15.6352	16.4790	17.5426

The MEANS procedure in Code 9.5 calculates confidence intervals using

$$\hat{V}_1(\tilde{\theta}) = \frac{1}{R}\frac{1}{R-1}\sum_{r=1}^{R}\left(\hat{\theta}_r - \tilde{\theta}\right)^2.$$

The confidence intervals in Output 9.5(b) are calculated using a t distribution with 6 (number of groups minus one) df.

Output 9.5(b). Calculate variances using the random group method (`example0904.sas`).

The MEANS Procedure

Variable	Label	Mean	Variance	Std Dev	Lower 95% CL for Mean	Upper 95% CL for Mean
Mean		16.6724103	0.1703510	0.4127360	16.2906932	17.0541274
Pctl_50	50 Percentile	15.9096822	0.1500055	0.3873055	15.5514844	16.2678801

The principles behind the replicate sample and random group methods underlie the replicate weight methods (balanced repeated replication, jackknife, and bootstrap) implemented in SAS software. These methods, described in Section 9.3, produce variance estimates that have more df (and hence are more stable) than random group methods, and thus are to be preferred for most situations.

9.3 Constructing Replicate Weights

Each of the survey analysis procedures in SAS software will create replicate weights for survey data, using any of the methods discussed in Chapter 9 of SDA. In this section, the replicate weights are created with the SURVEYMEANS procedure; the syntax for the other procedures (SURVEYFREQ, SURVEYREG, etc.) is the same. Lewis (2015) provides additional examples for using replicate weight methods in SAS software.

To create replicate weights with the SURVEYMEANS procedure, include the WEIGHT statement that specifies the survey weights, and, if applicable, include STRATA and CLUSTER statements that describe the survey design structure. Specify the VARMETHOD that you want in the PROC SURVEYMEANS statement. The options containing additional information go in parentheses behind the name of the variance estimation method. For replication methods, choose from:

VARMETHOD=BRR (REPS = numreps OUTWEIGHTS = myrepwgts) Creates balanced repeated replication (BRR) weights for designs with two psus per stratum. The user-supplied variable *numreps* contains the number of replications to be formed. If you omit the REPS= option, *numreps* is set to the smallest multiple of 4 that exceeds the number of strata. If you include the OUTWEIGHTS= option, the data set name specified (here called *myrepwgts*) will contain the original data as well as the *numreps* columns of replicate weights. If you specify REPS=24, for example, 24 replicate weight variables are created with names *RepWt_1, RepWt_2, ..., RepWt_24*. The BRR method also gives the option of using Fay's method (see Exercise 9.24 of SDA) with a specified value of ε.

VARMETHOD=JACKKNIFE (OUTWEIGHTS = myrepwgts OUTJKCOEFS = myjkcoefs) Creates replicate weights using the jackknife method by deleting one psu at a time. The data set named by the OUTWEIGHTS option contains the original data and columns of replicate weights named *RepWt_1, RepWt_2, ..., RepWt_K*, where K is the total number of psus in the data set. The data set created by the OUTJKCOEFS= option contains the jackknife coefficients.

VARMETHOD=BOOTSTRAP (REPS = numreps OUTWEIGHTS = myrepwgts SEED = myseed) Creates bootstrap weights. The user-supplied variable *numreps* contains the number of replicate weight variables to be created. The data set named by the OUTWEIGHTS option contains the original data and columns of replicate weights named *RepWt_1, RepWt_2, ..., RepWt_numreps.*

Each of the VARMETHOD options will store the replicate weights in a new data set specified by the OUTWEIGHTS option. Although you could create replicate weights for each analysis you want to carry out, it is more efficient to create them once and use the replicate weights in the new data set for subsequent analyses. This also ensures that, if a method such as bootstrap is used, where the replicate weights created from one SEED differ from those created with another SEED, all analyses done on the data set use the same set of replicate weights.

9.3.1 Balanced Repeated Replication

Example 9.5 of SDA. Code 9.6 shows the creation of BRR replicate weights for the small data set in Table 9.2 of SDA. Here, I assume that N = 10,000, so that variable *wt* contains the sampling weight, which is $N_h/2$ for this sample with 2 sampled observations per stratum.

Code 9.6. Create BRR weights for data in Table 9.2 of SDA (`example0905.sas`).

```
DATA brrex;
   INPUT strat strfrac y;
   wt = 10000*strfrac/2;
   DATALINES;
1 .3 2000
1 .3 1792
2 .1 4525
2 .1 4735
3 .05 9550
3 .05 14060
4 .1 800
4 .1 1250
5 .2 9300
5 .2 7264
6 .05 13286
6 .05 12840
7 .2 2106
7 .2 2070
;

PROC SURVEYMEANS DATA = brrex VARMETHOD=BRR(OUTWEIGHTS=repwts PRINTH) DF MEAN CLM;
  WEIGHT wt;
  STRATA strat;
  VAR y;
RUN;
```

The VARMETHOD=BRR option tells the SURVEYMEANS procedure to create BRR weights; I did not specify the REPS= option, so by default the procedure creates 8 replicate weights because that is the smallest multiple of 4 that is larger than 7, the number of strata.

Output 9.6(a). Create BRR weights for data in Table 9.2 of SDA (`example0905.sas`).

Data Summary	
Number of Strata	7
Number of Observations	14
Sum of Weights	10000

Variance Estimation	
Method	BRR
Number of Replicates	8

Statistics					
Variable	DF	Mean	Std Error of Mean	95% CL for Mean	
y	7	4451.700000	236.416465	3892.66389	5010.73611

The PRINTH option in the parentheses behind VARMETHOD asks the procedure to print the Hadamard matrix showing the balancing structure of the replicates. You usually do not need to see the Hadamard matrix, but I requested the option here so you can see the relation of the matrix in Output 9.6(b) to the matrix given in SDA. The BRR method in the SURVEYMEANS procedure includes the first psu of a stratum in the half-sample when $\alpha_{rh} = 1$, where the first psu is the psu that appears first in the data set. If you order the data differently, you may obtain a different set of replicate weights.

Output 9.6(b). Hadamard matrix for data in Table 9.2 of SDA (`example0905.sas`).

Hadamard Matrix						
Col1	Col2	Col3	Col4	Col5	Col6	Col7
1	1	1	1	1	1	1
-1	1	-1	1	-1	1	-1
1	-1	-1	1	1	-1	-1
-1	-1	1	1	-1	-1	1
1	1	1	-1	-1	-1	-1
-1	1	-1	-1	1	-1	1
1	-1	-1	-1	-1	1	1
-1	-1	1	-1	1	1	-1

Output 9.6(c) shows the first few replicate weights for this example. Each value in the replicate weight columns is either $2w_i$ or 0.

Output 9.6(c). Replicate weights for data in Table 9.2 of SDA (`example0905.sas`).

Obs	strat	strfrac	y	wt	RepWt_1	RepWt_2	RepWt_3	RepWt_4	RepWt_5	RepWt_6	RepWt_7	RepWt_8
1	1	0.30	2000	1500	3000	0	3000	0	3000	0	3000	0
2	1	0.30	1792	1500	0	3000	0	3000	0	3000	0	3000
3	2	0.10	4525	500	1000	1000	0	0	1000	1000	0	0
4	2	0.10	4735	500	0	0	1000	1000	0	0	1000	1000
5	3	0.05	9550	250	500	0	0	500	500	0	0	500
6	3	0.05	14060	250	0	500	500	0	0	500	500	0
7	4	0.10	800	500	1000	1000	1000	1000	0	0	0	0
8	4	0.10	1250	500	0	0	0	0	1000	1000	1000	1000
9	5	0.20	9300	1000	2000	0	2000	0	0	2000	0	2000
10	5	0.20	7264	1000	0	2000	0	2000	2000	0	2000	0
11	6	0.05	13286	250	500	500	0	0	0	0	500	500
12	6	0.05	12840	250	0	0	500	500	500	500	0	0
13	7	0.20	2106	1000	2000	0	0	2000	0	2000	2000	0
14	7	0.20	2070	1000	0	2000	2000	0	2000	0	0	2000

Fay's method for BRR. The SURVEYMEANS procedure also constructs replicate weights using Fay's method. Simply add the option FAY=epsilon, where epsilon is the desired value of ε, in the parentheses behind the BRR option. Code 9.7 uses $\varepsilon = 0.5$ for the data in Table 9.2 of SDA; in practice, values of ε between 0.3 and 0.5 work well.

Code 9.7. Create BRR weights using Fay's method (`example0905.sas`).

```
PROC SURVEYMEANS DATA = brrex VARMETHOD=BRR(OUTWEIGHTS=repwtsfay FAY=0.5) DF MEAN
    CLM;
  WEIGHT wt;
  STRATA strat;
  VAR y;
RUN;
```

Output 9.7(a). Create BRR weights using Fay's method (`example0905.sas`).

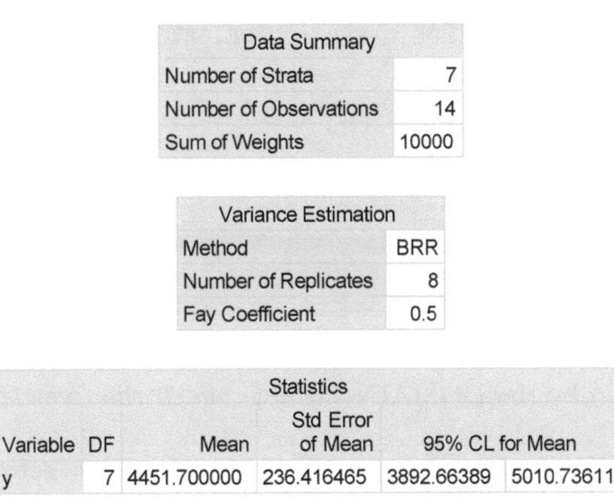

Data Summary	
Number of Strata	7
Number of Observations	14
Sum of Weights	10000

Variance Estimation	
Method	BRR
Number of Replicates	8
Fay Coefficient	0.5

Statistics					
Variable	DF	Mean	Std Error of Mean	95% CL for Mean	
y	7	4451.700000	236.416465	3892.66389	5010.73611

Each value in the replicate weight columns of Output 9.7(b) is either $\varepsilon w_i = 0.5w_i$, or $(2-\varepsilon)w_i = 1.5w_i$.

Output 9.7(b). Replicate weights using Fay's method for data in Table 9.2 of SDA (`example0905.sas`).

Obs	strat	strfrac	y	wt	RepWt_1	RepWt_2	RepWt_3	RepWt_4	RepWt_5	RepWt_6	RepWt_7	RepWt_8
1	1	0.30	2000	1500	2250	750	2250	750	2250	750	2250	750
2	1	0.30	1792	1500	750	2250	750	2250	750	2250	750	2250
3	2	0.10	4525	500	750	750	250	250	750	750	250	250
4	2	0.10	4735	500	250	250	750	750	250	250	750	750
5	3	0.05	9550	250	375	125	125	375	375	125	125	375
6	3	0.05	14060	250	125	375	375	125	125	375	375	125
7	4	0.10	800	500	750	750	750	750	250	250	250	250
8	4	0.10	1250	500	250	250	250	250	750	750	750	750
9	5	0.20	9300	1000	1500	500	1500	500	500	1500	500	1500
10	5	0.20	7264	1000	500	1500	500	1500	1500	500	1500	500
11	6	0.05	13286	250	375	375	125	125	125	125	375	375
12	6	0.05	12840	250	125	125	375	375	375	375	125	125
13	7	0.20	2106	1000	1500	500	500	1500	500	1500	1500	500
14	7	0.20	2070	1000	500	1500	1500	500	1500	500	500	1500

Example 9.6 of SDA. Now let's look at BRR for a data set with stratification and clustering: the NHANES data. Code 9.8 creates BRR weights for the NHANES data, using the pseudo-strata and pseudo-psus provided on the public-use data file. This data set contains 15 strata, so the procedure creates 16 replicate weights (the smallest multiple of 4 exceeding the number of strata) that are stored in variables *RepWt_1, RepWt_2, ..., RepWt_16*. The SURVEYMEANS statement in Code 7.4, which calculated linearization variances, contained the NOMCAR option to treat observations missing the value of *bmxbmi* as a domain; replication variance estimation methods automatically account for domains (see Section 9.5) and do not use the NOMCAR option.

Note that these replicate weights, constructed from the NHANES final weights available on the public-use data file, do not account for the effects of poststratification on the variances. See Section 9.3.4 for how to calculate replicate weights so that the variance estimates include the effects of weighting adjustments.

Code 9.8. Create BRR weights for the NHANES data (`example0906.sas`).

```
PROC SURVEYMEANS DATA=nhanes VARMETHOD = BRR (OUTWEIGHTS = nhanesBRR) DF MEAN CLM
    QUANTILE = (0.05 0.25 0.5 0.75 0.95);
  WEIGHT wtmec2yr;
  STRATA sdmvstra;
  CLUSTER sdmvpsu;
  DOMAIN age20;
  VAR bmxbmi;
run;
```

Output 9.8(a) shows the estimates for the entire sample; Output 9.8(b) gives the statistics for the two domains defined by the DOMAIN variable *age20* that was created in Code 7.3.

Output 9.8(a). Create BRR weights for the NHANES data. This set of statistics is from the VAR statement and includes members of all domains (`example0906.sas`).

The SURVEYMEANS Procedure

Data Summary	
Number of Strata	15
Number of Clusters	30
Number of Observations	9971
Number of Observations Used	9544
Number of Obs with Nonpositive Weights	427
Sum of Weights	316481044

Variance Estimation	
Method	BRR
Number of Replicates	16

Statistics

Variable	Label	DF	Mean	Std Error of Mean	95% CL for Mean	
bmxbmi	Body Mass Index (kg/m**2)	15	27.281654	0.202633	26.8497525	27.7135553

Quantiles

Variable	Label	Percentile		Estimate	Std Error	95% Confidence Limits	
bmxbmi	Body Mass Index (kg/m**2)	5		15.928655	0.105439	15.7039181	16.1533929
		25	Q1	21.920650	0.166847	21.5650243	22.2762764
		50	Median	26.509650	0.235259	26.0082067	27.0110924
		75	Q3	31.617098	0.252402	31.0791158	32.1550801
		95		41.265652	0.375849	40.4645493	42.0667542

Output 9.8(b). Domain statistics calculated with replicate weights (`example0906.sas`).

Statistics for age20 Domains

age20	Variable	Label	DF	Mean	Std Error of Mean	95% CL for Mean	
0	bmxbmi	Body Mass Index (kg/m**2)	15	20.602669	0.198384	20.1798228	21.0255145
1	bmxbmi	Body Mass Index (kg/m**2)	15	29.389101	0.259357	28.8362933	29.9419078

Quantiles for age20 Domains

age20	Variable	Label	Percentile		Estimate	Std Error	95% Confidence Limits	
0	bmxbmi	Body Mass Index (kg/m**2)	5		14.555706	0.068606	14.4094750	14.7019367
			25	Q1	16.244182	0.097587	16.0361801	16.4521835
			50	Median	18.988596	0.222628	18.5140762	19.4631167
			75	Q3	23.244709	0.270669	22.6677905	23.8216268
			95		31.943917	0.725194	30.3982025	33.4896318
1	bmxbmi	Body Mass Index (kg/m**2)	5		20.298934	0.207935	19.8557310	20.7421369
			25	Q1	24.353490	0.251236	23.8179925	24.8889873
			50	Median	28.234898	0.323854	27.5446196	28.9251758
			75	Q3	33.066151	0.304950	32.4161666	33.7161357
			95		42.640916	0.400948	41.7863156	43.4955155

The domain statistics of interest, for the domain of adults, are in the rows corresponding to *age20*=1 in Output 9.8(b). The mean and quantiles are estimated using the full sample weights, and the replicate weights are used to calculate the standard errors and confidence intervals. The estimated mean value of body mass index for adults is 29.389 with standard error 0.259, and the estimated median for adults is 28.235.

Note that the standard error for the median in Output 9.8(b), 0.323854, differs slightly from the value $0.3314 = \sqrt{0.1098}$ from the example in SDA. The SURVEYMEANS procedure smooths the quantiles (see Fuller, 2009, Section 4.2.3) before applying the replication variance estimation method. If desired, you can obtain the variance estimate given in SDA by typing VARMETHOD=BRR(NAIVEQVAR) in the SURVEYMEANS statement.

Mukhopadhyay et al. (2008) show how to implement a grouped balanced-half-sample method that can be used to construct BRR replicate weights for designs where some strata have more than two psus.

9.3.2 Jackknife

Jackknife weights for an SRS. Let's start by looking at how the SURVEYMEANS procedure creates jackknife weights and coefficients for an SRS, then move on to complex sample designs.

Example 9.7 of SDA. Output 9.2(b) shows the values of in-state and out-of-state tuition for the 10 colleges in replicate sample 1 (having *repgroup* = 1). First, for comparison, let's look at the linearization (Taylor series) estimate of the variance of the ratio of mean out-of-state tuition to mean in-state tuition, in Code and Output 9.9.

Code 9.9. Linearization variance estimates for an SRS of size 10 (`example0907.sas`).

```
DATA collegerep1;
   SET datalib.collegerg;
   IF repgroup = 1;    /* Keep the 10 observations in repgroup 1; delete the other
     repgroups */

PROC SURVEYMEANS DATA = collegerep1 VARMETHOD = TAYLOR;
   WEIGHT SamplingWeight;
   VAR tuitionfee_in tuitionfee_out;
   RATIO 'out-of-state to in-state' tuitionfee_out / tuitionfee_in;
RUN;
```

Output 9.9. Linearization variance estimates for an SRS of size 10 (`example0907.sas`).

Data Summary	
Number of Observations	10
Sum of Weights	500

Statistics						
Variable	Label	N	Mean	Std Error of Mean	95% CL for Mean	
TUITIONFEE_IN	TUITIONFEE_IN	10	8913.300000	454.457015	7885.2468	9941.3532
TUITIONFEE_OUT	TUITIONFEE_OUT	10	21615	2325.151140	16354.8427	26874.5573

Ratio Analysis: out-of-state to in-state						
Numerator	Denominator	N	Ratio	Std Error	95% CL for Ratio	
TUITIONFEE_OUT	TUITIONFEE_IN	10	2.424994	0.231178	1.90203408	2.94795414

Example 9.7 of SDA calculates the jackknife variance by omitting observation j in replicate j. This is equivalent to setting the weight of observation j to zero and distributing its weight among the other members of the SRS. Code 9.10 uses the SURVEYMEANS procedure to create jackknife weights for this SRS of size 10 and then calculates standard errors with the jackknife weights.

Code 9.10. Construct jackknife weights for an SRS of size 10 (`example0907.sas`).

```
PROC SURVEYMEANS DATA = collegerep1 VARMETHOD = JACKKNIFE (OUTJKCOEFS = jkcoefs
     OUTWEIGHTS = jkwgts);
   WEIGHT SamplingWeight;
   VAR tuitionfee_in tuitionfee_out;
   RATIO 'out-of-state to in-state' tuitionfee_out / tuitionfee_in;
RUN;
```

Output 9.10(a) shows the statistics produced by the SURVEYMEANS procedure. The jackknife output contains an extra box, not found in the linearization Output 9.9, informing that the procedure used the jackknife method with 10 replicates to estimate variances. The standard errors for the means and the ratio are calculated using the jackknife. The jackknife standard errors for the means are the same as for linearization because $\hat{V}_{\text{JK}}(\bar{y}) = s_y^2/n$ for an SRS, as shown in Section 9.3 of SDA. The jackknife standard error for the nonlinear statistic of the ratio, 0.231483, differs slightly from the linearization standard error of 0.231178. These values are extremely close—after all, the linearization variance and the jackknife variance are both consistent estimators for $V(\hat{B})$—but are not exactly the same.

Output 9.10(a). Construct jackknife weights for an SRS of size 10 in the SURVEYMEANS procedure (`example0907.sas`).

Data Summary	
Number of Observations	10
Sum of Weights	500

Variance Estimation	
Method	Jackknife
Number of Replicates	10

Statistics

Variable	Label	N	Mean	Std Error of Mean	95% CL for Mean	
TUITIONFEE_IN	TUITIONFEE_IN	10	8913.300000	454.457015	7885.2468	9941.3532
TUITIONFEE_OUT	TUITIONFEE_OUT	10	21615	2325.151140	16354.8427	26874.5573

Ratio Analysis: out-of-state to in-state

Numerator	Denominator	N	Ratio	Std Error	95% CL for Ratio	
TUITIONFEE_OUT	TUITIONFEE_IN	10	2.424994	0.231483	1.90134368	2.94864454

Output 9.10(b) prints the first two and last two jackknife weights created by the SURVEYMEANS procedure in the output data set *jkwgts*. The procedure assigns the names *RepWt_1* ... *RepWt_10* to the ten replicate weight variables it creates. Note that replicate weight variable *RepWt_j* assigns weight 0 to observation j and weight $55.5556 = w_i \times n/(n-1)$ to each of the other observations. Row j in Table 9.5 of SDA can be calculated using *RepWt_j* to calculate the means and ratio of means: $\bar{x}_{(j)} = \sum_{i \in \mathcal{S}} RepWt_{ji} x_i / \sum_{i \in \mathcal{S}} RepWt_{ji}$.

Output 9.10(b). Print jackknife replicate weights 1, 2, 9, and 10 from data set *jkwgts* (`example0907.sas`).

Obs	INSTNM	SamplingWeight	RepWt_1	RepWt_2	RepWt_9	RepWt_10
1	University of Louisiana at Lafayette	50	0.0000	55.5556	55.5556	55.5556
2	CUNY City College	50	55.5556	0.0000	55.5556	55.5556
3	Binghamton University	50	55.5556	55.5556	55.5556	55.5556
4	University of North Carolina at Chapel Hill	50	55.5556	55.5556	55.5556	55.5556
5	Oklahoma Panhandle State University	50	55.5556	55.5556	55.5556	55.5556
6	University of Science and Arts of Oklahoma	50	55.5556	55.5556	55.5556	55.5556
7	Rhode Island College	50	55.5556	55.5556	55.5556	55.5556
8	Francis Marion University	50	55.5556	55.5556	55.5556	55.5556
9	East Tennessee State University	50	55.5556	55.5556	0.0000	55.5556
10	Stephen F Austin State University	50	55.5556	55.5556	55.5556	0.0000

Output 9.10(c) prints the data set *jkcoefs*, where the variable *JKCoefficient* contains the jackknife coefficient of 0.9 for each observation. This is an SRS, so each jackknife coefficient equals $(n-1)/n$.

Output 9.10(c). Print jackknife coefficients from data set *jkcoefs* (`example0907.sas`).

Obs	Replicate	JKCoefficient
1	1	0.9
2	2	0.9
3	3	0.9
4	4	0.9
5	5	0.9
6	6	0.9
7	7	0.9
8	8	0.9
9	9	0.9
10	10	0.9

Now that the jackknife weights have been created in the SURVEYMEANS procedure (they could also have been created in any of the other survey analysis procedures), they can be used to estimate variances in any other survey procedure. Code 9.11 shows how to do that to estimate the regression relationship between in-state and out-of-state tuition for this sample. The only design features that need to be specified are in the WEIGHT and the REPWEIGHTS statements. Because this sample is so small, I overrode the default df calculations (which would use 10 df) to specify that $(n - 2) = 8$ df should be used; with large samples, you can just trust the survey analysis procedures to calculate the df.

Code 9.11. Use the jackknife weights to estimate a regression line (`example0907.sas`).

```
PROC SURVEYREG DATA = jkwgts VARMETHOD = JACKKNIFE;
   WEIGHT SamplingWeight;
   REPWEIGHTS RepWt_1 - RepWt_10 / JKCOEFS = jkcoefs;
   MODEL tuitionfee_out = tuitionfee_in / SOLUTION DF=8;
RUN;
```

Output 9.11. Use the jackknife weights to estimate a regression line (`example0907.sas`).

Estimated Regression Coefficients

Parameter	Estimate	Standard Error	t Value	Pr > |t|
Intercept	481.281219	22346.2733	0.02	0.9833
TUITIONFEE_IN	2.370998	2.6286	0.90	0.3934

Note: The degrees of freedom for the t tests is 8.

Jackknife weights for a complex survey. For a survey with stratification and clustering, jackknife weights are constructed similarly. The only difference is that you need to include the STRATA and CLUSTER statements in the SURVEYMEANS procedure. This guarantees that the procedure will keep observations in the same cluster together when creating the replicate weights.

Example 9.8 of SDA. Code 5.3 shows calculations using the SURVEYMEANS procedure with linearization variance estimates. Now let's construct the jackknife weights in Code 9.12. Because the CLUSTER statement is included, the procedure deletes one psu at a time rather than one observation at a time.

Code 9.12. Create jackknife weights for the coots data (`example0908.sas`).

```
PROC SURVEYMEANS DATA = coots VARMETHOD = JACKKNIFE (OUTWEIGHTS=jkwt OUTJKCOEFS=
    jkcoef);
   WEIGHT relwt;
   CLUSTER clutch;
   VAR volume;
RUN;
```

Output 9.12(a). Create jackknife weights for the coots data (`example0908.sas`).

The SURVEYMEANS Procedure

Data Summary	
Number of Clusters	184
Number of Observations	368
Sum of Weights	1758

Variance Estimation	
Method	Jackknife
Number of Replicates	184

Statistics					
Variable	N	Mean	Std Error of Mean	95% CL for Mean	
volume	368	2.490778	0.061036	2.37035364	2.61120281

Output 9.12(b) prints the first six replicate weight values for the first ten observations. *RepWt_1* has weight 0 for both observations in psu 1, *RepWt_2* has weight 0 for both observations in psu 2, and so on. The SURVEYMEANS procedure sets the replicate weight variable equal to 0 for all observations from the psu being deleted in that replicate.

Output 9.12(b). Print the first 6 jackknife weights for the coots data (`example0908.sas`).

Obs	clutch	csize	length	breadth	volume	tmt	relwt	RepWt_1	RepWt_2	RepWt_3	RepWt_4	RepWt_5	RepWt_6
1	1	13	44.30	31.10	3.79576	1	6.5	0.00000	6.53552	6.53552	6.53552	6.53552	6.53552
2	1	13	45.90	32.70	3.93285	1	6.5	0.00000	6.53552	6.53552	6.53552	6.53552	6.53552
3	2	13	49.20	34.40	4.21560	1	6.5	6.53552	0.00000	6.53552	6.53552	6.53552	6.53552
4	2	13	48.70	32.70	4.17276	1	6.5	6.53552	0.00000	6.53552	6.53552	6.53552	6.53552
5	3	6	51.05	34.25	0.93176	0	3.0	3.01639	3.01639	0.00000	3.01639	3.01639	3.01639
6	3	6	49.35	34.40	0.90074	0	3.0	3.01639	3.01639	0.00000	3.01639	3.01639	3.01639
7	4	11	49.20	31.55	3.01827	1	5.5	5.53005	5.53005	5.53005	0.00000	5.53005	5.53005
8	4	11	48.55	33.10	2.97840	1	5.5	5.53005	5.53005	5.53005	0.00000	5.53005	5.53005
9	5	10	49.40	34.55	2.50458	1	5.0	5.02732	5.02732	5.02732	5.02732	0.00000	5.02732
10	5	10	49.05	34.95	2.48684	1	5.0	5.02732	5.02732	5.02732	5.02732	0.00000	5.02732

Example 9.8 of SDA has one stratum containing 184 psus. With a stratified multistage sample, include a STRATA statement as well as the CLUSTER statement; the survey analysis procedure will then construct replicate weights by deleting one psu at a time from each stratum.

9.3.3 Bootstrap

Example 9.9 of SDA. The survey analysis procedures, such as the SURVEYMEANS procedure, create bootstrap weights using the method in Rao et al. (1992). Code 9.13 shows the creation of 1000 bootstrap weights to estimate the population distribution of height in the SURVEYMEANS procedure for the SRS of heights in data *htsrs*. The weight variable *wt* equals 10 for each observation.

Code 9.13. Create bootstrap weights for SRS of heights (`example0909.sas`).

```
PROC SURVEYMEANS DATA = htsrs VARMETHOD = BOOTSTRAP (REPS = 1000 OUTWEIGHTS =
    bootwts SEED=735490981) MEAN CLM DF PERCENTILE = (25 50 75);
   WEIGHT wt;
   VAR height;

PROC PRINT DATA = bootwts (OBS = 8);
   VAR height wt RepWt_1-RepWt_6;
RUN;
```

Output 9.13(a) shows the summary statistics that are calculated using the bootstrap.

Output 9.13(a). Create bootstrap weights for the SRS of heights (`example0909.sas`).

The SURVEYMEANS Procedure

Data Summary	
Number of Observations	200
Sum of Weights	2000

Variance Estimation	
Method	Bootstrap
Number of Replicates	1000
Bootstrap Seed	735490981

Statistics					
Variable	DF	Mean	Std Error of Mean	95% CL for Mean	
height	199	168.940000	0.780060	167.401756	170.478244

Quantiles						
Variable	Percentile		Estimate	Std Error	95% Confidence Limits	
height	25	Q1	159.700000	1.091987	157.546649	161.853351
	50	Median	168.750000	0.940516	166.895342	170.604658
	75	Q3	176.000000	0.914887	174.195883	177.804117

Output 9.13(b) prints the first 8 observations and the first six replicate weight variables created in data set *bootwts*. In replicate 1, observation 1 is selected four times; the weight in the replicate for that observation is, using the formula in Section 9.3 of SDA, $10 \times (200/199) \times 4 = 40.201$.

Output 9.13(b). Print the first 8 observations from bootstrap weight data set (`example0909.sas`).

Obs	height	wt	RepWt_1	RepWt_2	RepWt_3	RepWt_4	RepWt_5	RepWt_6
1	159	10	40.2010	0.0000	0.0000	0.0000	20.1005	40.2010
2	174	10	10.0503	10.0503	0.0000	10.0503	20.1005	0.0000
3	186	10	20.1005	20.1005	10.0503	0.0000	0.0000	30.1508
4	158	10	0.0000	0.0000	10.0503	0.0000	0.0000	10.0503
5	178	10	0.0000	0.0000	50.2513	0.0000	10.0503	10.0503
6	177	10	10.0503	10.0503	0.0000	10.0503	10.0503	10.0503
7	168	10	10.0503	10.0503	0.0000	0.0000	20.1005	0.0000
8	159	10	30.1508	20.1005	10.0503	20.1005	0.0000	20.1005

The SURVEYMEANS procedure's bootstrap for an SRS differs slightly from the bootstrap procedure that is usually presented in the model-based setting (Singh and Xie, 2010), in which with-replacement samples of size n are repeatedly sampled from the original SRS of size n. That bootstrap is also easy to carry out in SAS software, using the SURVEYSELECT procedure; see Code 12.5 for an example.

For an SRS, the SURVEYMEANS procedure takes repeated with-replacement samples of size $n-1$; the weights are multiplied by $n/(n-1)$ so that the sum of weights for each replicate is N. For large n, the results from the two bootstrap procedures (taking samples of size n or taking samples of size $n-1$) are practically identical. A stratified multistage sample, however, may have only 2 psus in some strata, so that sampling $n-1$ psus instead of n psus makes a difference and the SURVEYMEANS procedure should be used to create bootstrap weights for complex samples.

Example 9.10 of SDA. The code for creating bootstrap weights in a complex survey design is similar to that for an SRS. The SURVEYMEANS (or other survey analysis) procedure is used to create the weights, and the survey design is specified by the STRATA and CLUSTER statements. The data set *htstrat* in Example 9.10 is a stratified random sample, so Code 9.14 requires only the STRATA statement to define the survey design. Output 9.14 shows the weight variable *wt* and the first 6 bootstrap replicate weights (of the 1000 replicate weight variables created) for the first 8 observations. If you use a different SEED, you will get a different set of bootstrap replicates and slightly different standard errors.

Code 9.14. Create bootstrap weights for stratified random sample of heights (`example0910.sas`).

```
PROC SURVEYMEANS DATA = htstrat VARMETHOD = BOOTSTRAP (REPS = 1000 OUTWEIGHTS =
    strbootwts SEED=982537455) MEAN CLM DF PERCENTILE = (0 25 50 75 100);
    WEIGHT wt;
    STRATA gender;
    VAR height;

PROC PRINT DATA = strbootwts (OBS = 8);
   VAR height gender wt RepWt_1-RepWt_6;
RUN;
```

Output 9.14. Print the first 8 bootstrap weights for the stratified sample of heights (`example0910.sas`).

Obs	height	gender	wt	RepWt_1	RepWt_2	RepWt_3	RepWt_4	RepWt_5	RepWt_6
1	166	F	6.25	12.5786	6.2893	0.0000	6.2893	6.2893	6.2893
2	163	F	6.25	18.8679	0.0000	0.0000	0.0000	6.2893	0.0000
3	166	F	6.25	6.2893	0.0000	12.5786	18.8679	6.2893	6.2893
4	155	F	6.25	12.5786	0.0000	0.0000	0.0000	6.2893	12.5786
5	154	F	6.25	6.2893	6.2893	6.2893	6.2893	6.2893	6.2893
6	160	F	6.25	6.2893	0.0000	6.2893	18.8679	12.5786	0.0000
7	170	F	6.25	6.2893	0.0000	6.2893	0.0000	0.0000	25.1572
8	156	F	6.25	0.0000	12.5786	6.2893	0.0000	12.5786	18.8679

9.3.4 Replicate Weights and Nonresponse Adjustments

The code given so far in this section constructs replicate sampling weights. When nonresponse adjustments are made to the final weights, as described in Chapter 8 of SDA, the steps of weighting class adjustments, poststratification, raking, and other adjustments that are used on the final weights need to be repeated for each replicate weight column.

The SURVEYMEANS procedure will create replicate weights that reflect poststratification weight adjustments when the POSTSTRATA statement is used along with VARMETHOD= BRR, JACKKNIFE, or BOOTSTRAP. Let's look at that for the poststratified weights in Example 4.9 of SDA, which was discussed in Section 4.4.

Example 4.9 of SDA. Code 4.8 created the data set *pstot* containing the poststratification totals. These are used in the SURVEYMEANS procedure in Code 9.15 to create 500 bootstrap replicate weights. The output data set *agsrsboot* contains the poststratified weights in variable *_PSWt_* as well as the 500 replicate weights *RepWt_1–RepWt_500*, each of which has been poststratified to the totals in data set *pstot*. This data set is an SRS; for other designs, include the STRATA and CLUSTER statements when creating the replicate weights.

Code 9.15. Creating poststratified bootstrap weights for `agsrs.csv` (`example0409.sas`).

```
PROC SURVEYMEANS DATA = agsrs VARMETHOD = BOOTSTRAP (REPS = 500 OUTWEIGHTS =
    agsrsboot SEED = 38373456);
  WEIGHT sampwt;
  VAR acres92;
  POSTSTRATA region / PSTOTAL = pstot;
RUN;
```

If you check the sum of the poststratified weights *_PSWt_* and poststratified replicate weights *RepWt_1–RepWt_500* for the variable *region*, you will find that each sums to 1054 for the NC region, to 220 for the NE region, to 1382 for the S region, and to 422 for the W region—exactly the poststratification totals defined in the data set *pstot* in Code 4.8.

The estimated bootstrap variance of the counts in the poststrata (variable *region*) is therefore zero (well, it is zero up to roundoff error), as shown in Code and Output 9.16. The confidence intervals for variable *acres92* account for the poststratification, but do not use a finite population correction since no TOTAL= option is requested.

Code 9.16. Estimates with poststratified bootstrap weights (`example0409.sas`).

```
PROC SURVEYMEANS DATA = agsrsboot VARMETHOD = BOOTSTRAP MEAN CLM SUM CLSUM;
   CLASS region;
   WEIGHT _PSWt_;
   REPWEIGHTS RepWt_1 - RepWt_500;
   VAR region acres92;
RUN;
```

Output 9.16. Estimates with poststratified bootstrap weights (`example0409.sas`).

Variance Estimation	
Method	Bootstrap
Replicate Weights	AGSRSBOOT
Number of Replicates	500

Statistics									
Variable	Level	Mean	Std Error of Mean	95% CL for Mean		Sum	Std Error of Sum	95% CL for Sum	
acres92		299778	18189	264042.350	335513.876	922717031	55984918	812722352	1032711710
region	NC	0.342430	0	0.342	0.342	1054.000000	7.275247E-13	1054	1054
	NE	0.071475	0	0.071	0.071	220.000000	5.053952E-14	220	220
	S	0.448993	0	0.449	0.449	1382.000000	2.944801E-12	1382	1382
	W	0.137102	0	0.137	0.137	422.000000	1.351631E-13	422	422

The SURVEYMEANS procedure is limited to creating replicate weights for simple poststratification. If you have multiple steps of nonresponse adjustments before poststratification, you will need to use a custom-written program to create the final and replicate weights.

Performing multiple steps of nonresponse adjustments. Many surveys have several steps of weighting class adjustments followed by calibration; sometimes intermediate or final weights are trimmed or smoothed so that the weight adjustments do not have "spikes" for some observations. Each step must be repeated for each replicate sampling weight.

Thus, if weighting class adjustments are used, the steps in Section 9.3.4 of SDA must be carried out for each replicate r, for $r = 1$ to R. Suppose there are C weighting classes. Replicate weights that account for the weighting class adjustments are created as follows.

1. Start with the sampling weight vector $\mathbf{w}$, which is calculated as the inverse of the probability of selection for each member of the selected sample, whether respondent or nonrespondent.

2. Create R replicate sampling weights $\mathbf{w}_1, \ldots, \mathbf{w}_R$ using the desired replication variance estimation method; this can be done in any of the survey analysis procedures in SAS software. At this stage, the data set contains records for everyone in the selected sample.

3. Now create weighting-class-adjusted weight vector $\tilde{\mathbf{w}}$, where $\tilde{w}_i = w_i \times$ (sum of weights in $\mathbf{w}$ for selected sample in class c) / (sum of weights in $\mathbf{w}$ for respondents in class c) if unit i is a respondent and $\tilde{w}_i = 0$ if unit i is a nonrespondent. This transfers the weights of the nonrespondents in weighting class c to the respondents in that class.

4. Repeat the operation in Step 3 for each column of replicate weights. Create modified replicate weight variable $\tilde{\mathbf{w}}_r$ by multiplying each element by the appropriate weighting factor, calculated using replicate weight r: for observation i in class c, the modified replicate weight is $\tilde{w}_{ir} = w_{ir} \times$ (sum of weights in $\mathbf{w}_r$ for selected sample in class c) / (sum of weights in $\mathbf{w}_r$ for respondents in class c).

After the weighting class adjustments, the next step of the weighting needs to be applied to the full-sample adjusted weight $\tilde{\mathbf{w}}$ and to each weighting-class-adjusted replicate weight variable $\tilde{\mathbf{w}}_r$. Thus if the weights are poststratified after carrying out the weighting class adjustment, the poststratification needs to be done on $\tilde{\mathbf{w}}$ and then on each $\tilde{\mathbf{w}}_r$. (Note that the weighting class adjustment differs from poststratification: for poststratification, the control totals, obtained from an external source, are the same for each replicate weight while for weighting class adjustments, the sum of the weights in $\mathbf{w}_r$ for the selected sample in class c depends on the weights in replicate r.)

Each additional step in the weighting adjustments needs to be carried out separately on each replicate weight. Some of the macros for nonresponse adjustment referenced in Section 8.2 will create replicate weights for the nonresponse adjustments they carry out. In general, however, with complicated weighting adjustments, the survey producer will need to customize the steps and write a survey-specific program to carry out all the steps of nonresponse adjustments on the replicate sampling weights and compute the set of final replicate weights.

9.4 Computing Estimates with Replicate Weights

We have already seen how to use the survey analysis procedures in SAS software to calculate estimates with replicate weights that were created in the software. Code 9.11 used jackknife replicate weights to conduct a regression analysis, and Code 9.16 illustrated the use of bootstrap replicate weights.

The general structure is similar for using any type of replicate weights, including those made available on public-use data files by organizations that collected the data and created the weights. Code 9.17 gives the basic form for calculating estimates with replicate weights. These are shown for the SURVEYMEANS procedure, but any other survey analysis procedure follows the same format.

Code 9.17. Calculate estimates with replicate weights.

```
/* BRR replicate weights in data set brrwtdata */
PROC SURVEYMEANS DATA = brrwtdata VARMETHOD = BRR (FAY = fay_coefficient)
    statistics-keywords;
   WEIGHT finalwt;
   REPWEIGHTS RepWt_1 - RepWt_R;
   VAR variables;

/* Bootstrap replicate weights in data set bootwtdata */
PROC SURVEYMEANS DATA = bootwtdata VARMETHOD = BOOTSTRAP statistics-keywords;
   WEIGHT finalwt;
   REPWEIGHTS RepWt_1 - RepWt_R;
   VAR variables;

/* Jackknife replicate weights in data set jkwtdata, coefficients in myjkcoef */
PROC SURVEYMEANS DATA = jkwtdata VARMETHOD = JACKKNIFE statistics-keywords;
   WEIGHT finalwt;
   REPWEIGHTS RepWt_1 - RepWt_R / JKCOEFS = myjkcoef;
   VAR variables;
```

To compute estimates with already-created replicate weights, specify the VARMETHOD and include WEIGHT and REPWEIGHTS statements; these are needed for all replicate weight methods. Because the stratification and clustering of the survey design have been accounted for when the replicate weights were constructed, do not include STRATA or CLUSTER statements. For BRR, if Fay's method has been used, specify the Fay coefficient as an option to the method. For jackknife, specify the jackknife coefficients either as a constant (if all replicates have the same coefficient) or as a data set, where the variable named *JKCoefficient* contains the coefficients for all of the replicate weight variables.

Some data sets contain replicate weights that are created using a variant of BRR, jackknife, or bootstrap. You can calculate variance estimates using any of these variants in the survey analysis procedures of SAS software. The general form of a replication variance estimator is given in Equation (9.1):

$$\hat{V}_{\text{rep}}(\hat{\theta}) = \sum_{r=1}^{R} c_r(\hat{\theta}_r - \hat{\theta})^2. \tag{9.1}$$

The VARMETHOD=JACKKNIFE option can be used to calculate standard errors for any type of replicate weights that have a variance estimate in the form of (9.1). Simply list the variables containing the replicate weights in the REPWEIGHTS statement, and include the coefficients c_r in the JKCOEFS option.

A data producer usually provides a report that describes the type of replicate weights that are produced and gives the coefficients c_r. For example, the documentation for the American Community Survey (U.S. Census Bureau, 2020) states that the 80 replicate weights provided for the public-use data files are to be used with coefficients $c_r = 4/80$. You can then use the basic jackknife code in Code 9.17 to analyze the data, specifying that JKCOEFS = 0.05 or including a data set *jkcoef_ data* that contains the value 0.05 for each of the 80 observations in the variable named *JKCoefficient*.

Statistics calculated in non-survey procedures of SAS software. The replicate weights can also be used to calculate standard errors for statistics that are not calculated in SAS software's survey procedures but can be calculated with a non-survey procedure. For example, the GENMOD procedure fits generalized linear models to data sets but does not account for stratification and clustering. If one wants to fit a Poisson regression, however, the procedure allows estimates to be calculated using the survey weights. This gives the correct point estimates for the regression parameters, but with incorrect standard errors that do not account for the survey design. The correct standard errors can be calculated by rerunning the GENMOD procedure with each set of replicate weights, then using (9.1) to calculate the variance of each regression parameter. SAS Institute Inc. (2015) provides a macro for doing this. Section 10.3 and Appendix B of this book show how to use jackknife weights to calculate the standard errors of loglinear model coefficients.

Before using a non-survey procedure with replicate weights, however, make sure that the procedure treats the variable entered in the WEIGHT statement as a sampling weight—that is, the procedure computes estimates as though there are w_i copies of observation i. For the GENMOD procedure, for example, the WEIGHT variable is the exponential family dispersion parameter weight. The GENMOD procedure produces survey-weighted estimates for Poisson regression, but does not produce survey-weighted estimates for negative binomial regression. Section B.1 discusses how weights are used in the GENMOD procedure and describes how you can tell which non-survey procedures in SAS software are amenable to replicate weight calculations.

9.5 Domain Estimates with Replicate Weights

Section 4.3 warned that you must use the DOMAIN statement when calculating estimates for domains. The survey analysis procedures, when used with linearization variance estimates, all give a warning in the log if you use a BY or WHERE statement to analyze a subset of the data:

```
NOTE: The BY statement provides completely separate analyses of the BY groups.
      It does not provide a statistically valid subpopulation or domain
      analysis, where the total number of units in the subpopulation is not
      known with certainty. If you want a domain analysis, you should include
      the DOMAIN variables in a DOMAIN statement.
```

The SURVEYMEANS procedure gives this warning because the sample size for each domain within a psu is actually a random variable; if we drew a different sample, we would obtain a different number of observations in the psu that fall within a specific subpopulation. Thus, to estimate the mean value of the body mass index for adults age 20 and over, as in Example 7.9 of SDA and Code 7.4, the DOMAIN statement is needed to account for the sample-to-sample variability in the number of adults in each psu, and for the possibility that some psus might contain no sample members in the domain of interest (unlikely to occur for the domain of all adults, but this can occur for smaller domains such as the domain of persons who are medical doctors).

This warning, however, applies only to variances calculated using the linearization method, and the SURVEYMEANS procedure does not produce it if you use the BY or WHERE statement along with a REPWEIGHTS statement. If you are using replication methods to estimate variances, you can analyze subsets of the data—provided that the replicate weights have been created using the full sample.

Why does this work? Look at the general form of the replication variance estimator in (9.1). When you calculate domain estimates, $\hat{\theta}$ and the replicate values $\hat{\theta}_r$ are calculated using the same subset of observations. Suppose that θ is the population mean for variable y in domain d, say the proportion of medical doctors who have high blood pressure. Let $x_i = 1$ if observation i is in domain d (is a medical doctor) and 0 otherwise. Then

$$\hat{\theta} = \frac{\sum_{i \in \mathcal{S}} w_i x_i y_i}{\sum_{i \in \mathcal{S}} w_i x_i} = \frac{\sum_{i \in \mathcal{S}_d} w_i y_i}{\sum_{i \in \mathcal{S}_d} w_i},$$

where $\mathcal{S}_d$ is the subset of observations in domain d. The estimated domain mean can be calculated using only the subset of observations in the domain. The same is true for each column of replicate weights, with $\hat{\theta}_r = \sum_{i \in \mathcal{S}_d} w_{ir} y_i / \sum_{i \in \mathcal{S}_d} w_{ir}$. Thus, each term $c_r(\hat{\theta}_r - \hat{\theta})^2$ in (9.1) is the same whether the full sample is used or whether $\hat{\theta}_r$ and $\hat{\theta}$ are calculated using only the subset of observations in domain d.

The replicate weights account for the variability in the domain sample sizes across psus. You can, of course, still use the DOMAIN statement in the survey analysis procedure along with the replicate weights, but for some analyses, it may be easier to just form a subset of the data, or to use a BY or WHERE statement to analyze the subset. Make sure, though, that the replicate weights and the coefficients c_r are created using the whole data set, not just for the subset you wish to analyze.

9.6 Variance Estimation for Quantiles

Section 9.5 of SDA gives a confidence interval for the qth quantile, θ_q, using the Woodruff (1952) method, as

$$\left[\hat{F}^{-1}\left\{q - 1.96\sqrt{\hat{V}[\hat{F}(\hat{\theta}_q)]}\right\}, \hat{F}^{-1}\left\{q + 1.96\sqrt{\hat{V}[\hat{F}(\hat{\theta}_q)]}\right\}\right], \tag{9.2}$$

where $\hat{F}$ is the empirical cumulative distribution function (cdf) estimated from the survey.

The SURVEYMEANS procedure computes confidence intervals for quantiles with a slightly different formula, as

$$\left[\hat{F}^{-1}\left\{\hat{F}(\hat{\theta}_q) - t_{\text{df}}\sqrt{\hat{V}[\hat{F}(\hat{\theta}_q)]}\right\}, \hat{F}^{-1}\left\{\hat{F}(\hat{\theta}_q) + t_{\text{df}}\sqrt{\hat{V}[\hat{F}(\hat{\theta}_q)]}\right\}\right].$$

Here, $\hat{F}$ is the interpolated empirical cdf from Section 7.2 and t_{df} is the critical value from a t distribution with df degrees of freedom. This interval is asymptotically equivalent to the confidence interval in (9.2) when it is assumed the empirical cdf approaches a continuous function as the sample size increases, since then $\hat{F}(\hat{\theta}_q)$ is expected to be close to q. The two intervals may differ, however, for samples where y takes on a limited number of values.

Example 9.12 of SDA. Code 9.18 calculates the specified percentiles for the data set *htstrat*. No standard error is given for the minimum (PERCENTILE = 0) and maximum (PERCENTILE = 100) because the conditions for the Woodruff method are not met for these values of q.

Code 9.18. Confidence intervals for quantiles (`example0912.sas`).

```
PROC SURVEYMEANS DATA = htstrat MEAN CLM PERCENTILE=(0 25 50 75 100);
    WEIGHT wt;
    STRATA gender;
    VAR height;
RUN;
```

Output 9.18. Confidence intervals for quantiles (`example0912.sas`).

Quantiles						
Variable	Percentile		Estimate	Std Error	95% Confidence Limits	
height	0	Min	142.000000	.	.	.
	25	Q1	160.714286	0.766758	159.202226	162.226346
	50	Median	167.555556	1.042272	165.500177	169.610934
	75	Q3	176.625000	1.353653	173.955572	179.294428
	100	Max	200.000000	.	.	.

The standard errors and confidence limits in Output 9.18 differ slightly from those given in Example 9.12 of SDA because a different method is used for calculation. The confidence intervals in Output 9.18 are symmetric; option NONSYMCL will produce asymmetric confidence intervals.

Standard errors for quantiles using replication variance estimation methods. The SURVEYMEANS procedure will also compute standard errors and confidence intervals for quantiles when replicate weights are supplied. We saw this in Code 9.13 with the bootstrap.

The variance estimator of the form in (9.1) is not appropriate for nonsmooth functions such as quantiles when the jackknife is used—if you substitute the estimated quantiles for $\hat{\theta}$ and $\hat{\theta}_r$ into (9.1), the variance estimator is not consistent. But the jackknife can be used with a smoothed version of the quantile function. The SURVEYMEANS procedure uses a smoothing method similar to that in Section 4.2.3 of Fuller (2009) to construct confidence intervals for quantiles when replication variance estimates are used.

9.7 Summary, Tips, and Warnings

The survey analysis procedures in SAS software provide multiple options for estimating variances. They will estimate variances using linearization methods for commonly used statistics such as means, proportions, ratios, and regression coefficients.

In addition, the procedures will create replicate weights using the BRR, jackknife, and bootstrap methods. They will also analyze data sets in which the replicate weights have been created by someone else. This allows variances to be estimated for almost any statistic that is a smooth function of population totals, whether that statistic has been specifically included in the survey analysis procedures or not. The BRR and bootstrap methods (and some variants of the jackknife method) also calculate variances for some statistics that are not smooth functions of population totals, such as quantiles.

Tips and Warnings

- To create replicate weights for a complex survey in one of the survey analysis procedures such as the SURVEYMEANS procedure, include the WEIGHT, STRATA, and CLUSTER statements to specify the design. Then the procedure will create the weights using the method you specify in the VARMETHOD= option. Check that the sum of the weights output by the procedure approximately equals the population size.
- To analyze data for which replicate weights have already been created, include the WEIGHT and REPWEIGHTS statements. Do not include STRATA or CLUSTER statements when you analyze data using replicate weights, because the design information has already been incorporated into the replicate weights.
- For domain estimation when linearization variance estimation is used, include the DOMAIN statement. When using replication variance estimation, you can find estimates for domains either by including the DOMAIN statement or by creating a subset of the data with the domain of interest. Make sure, though, that the replicate weights are created using the full data set, not just with the subset of interest.

10

Categorical Data Analysis in Complex Surveys

The SURVEYFREQ procedure performs categorical data analyses on survey data. Code 2.6 and 3.14 showed how to use the SURVEYFREQ procedure to calculate percentages in a one-way classification, and Code 7.5 showed how to estimate domain percentages. Now let's explore how to use the SURVEYFREQ procedure to produce multi-way contingency tables, estimate odds ratios and other measures of association, and conduct chi-square tests of independence.

10.1 Contingency Tables and Odds Ratios

First let's look at the contingency table and odds ratio for a simple random sample (SRS).

Example 10.1 of SDA. Data set *cable1* contains the counts from each category formed by the cross-classification of *computer* and *cable*. This form could be used to perform a chi-square test in the non-survey FREQ procedure, but to perform an analysis in the SURVEYFREQ procedure, the data must be converted to a form where each record gives the *computer* (y or n) and *cable* (y or n) status for one household, along with the weight (here, set to the relative weight of 1) for that household. Code 10.1 creates data set *cable2*, which has 500 records, and produces the contingency table shown in Example 10.1 of SDA.

Code 10.1. Contingency table and odds ratio for SRS (`example1001.sas`).

```
DATA cable1;
   INPUT computer $ cable $ count;
   DATALINES;
n n      105
n y      188
y n       88
y y      119
;
DATA cable2;
   SET cable1;
   DO i = 1 TO count;
      wt = 1;
      OUTPUT;
   END;

PROC SURVEYFREQ DATA=cable2;
   WEIGHT wt;
   TABLES cable*computer / CHISQ LRCHISQ EXPECTED OR CL;
RUN;
```

DOI: 10.1201/9781003160366-10

Output 10.1(a). Contingency table for SRS (`example1001.sas`).

Table of cable by computer									
cable	computer	Frequency	Weighted Frequency	Std Err of Wgt Freq	Expected Wgt Freq	Percent	Std Err of Percent	95% Confidence Limits for Percent	
n	n	105	105.00000	9.11681	113.09800	21.0000	1.8234	17.4176	24.5824
	y	88	88.00000	8.52393	79.90200	17.6000	1.7048	14.2506	20.9494
	Total	193	193.00000	10.89676		38.6000	2.1794	34.3182	42.8818
y	n	188	188.00000	10.84191	179.90200	37.6000	2.1684	33.3397	41.8603
	y	119	119.00000	9.53204	127.09800	23.8000	1.9064	20.0544	27.5456
	Total	307	307.00000	10.89676		61.4000	2.1794	57.1182	65.6818
Total	n	293	293.00000	11.02475		58.6000	2.2049	54.2679	62.9321
	y	207	207.00000	11.02475		41.4000	2.2049	37.0679	45.7321
	Total	500	500.00000	0		100.0000			

The contingency table provided by the SURVEYFREQ procedure in Output 10.1(a) is in "list" form, with the row variable levels listed in Column 1 and the column variable levels listed in Column 2. The contingency table margins are designated by "Total." To get to the cross-classified form given in Table 10.1 of SDA, you need to enter the numbers from the "Weighted Frequency" column into the cells specified by the levels of the *cable* and *computer* variables in the first two columns of the output. The table of expected weighted counts, under the null hypothesis that the row and column variables are independent, is formed similarly from the numbers in the "Expected Wgt Freq" column (produced when the EXPECTED option is included in the TABLES statement).

The CL option in the TABLES statement requests confidence intervals for the estimated percentages in the cross-classified cells and margins. If desired, you could also include the CLWT option, which requests confidence intervals for the weighted frequencies. Other options for the TABLES statement are listed in Table 10.2 at the end of this chapter.

The OR option in the TABLES statement requests the odds ratio, shown in Output 10.1(b). The SURVEYFREQ procedure calculates confidence limits for the odds ratio by exponentiating the confidence limits for the log odds ratio discussed in Example 10.1 of SDA: $[\exp(-0.646), \exp(0.084)] = [0.524, 1.089]$.

Output 10.1(b). Odds ratio for SRS (`example1001.sas`).

Odds Ratio and Relative Risks (Row1/Row2)			
Statistic	Estimate	95% Confidence Limits	
Odds Ratio	0.7553	0.5238	1.0889
Column 1 Relative Risk	0.8884	0.7590	1.0398
Column 2 Relative Risk	1.1763	0.9541	1.4502
Sample Size = 500			

The CHISQ and LRCHISQ options in the TABLES statement of Code 10.1 request the Pearson and likelihood ratio chi-square statistics. Because this is an SRS, the SURVEYFREQ procedure gives the test statistics calculated under multinomial sampling assumptions, with $X^2 = 2.281$ and $G^2 = 2.275$, and these are the same values that would be calculated if the FREQ procedure were used. Note that the Design Correction for each test equals 1 because the sample is an SRS. Section 10.2 will show how chi-square test statistics are calculated for data from a complex survey.

Output 10.1(c). Pearson and likelihood ratio chi-square tests for SRS (`example1001.sas`).

Rao-Scott Chi-Square Test	
Pearson Chi-Square	2.2810
Design Correction	1.0000
Rao-Scott Chi-Square	2.2810
DF	1
Pr > ChiSq	0.1310
F Value	2.2810
Num DF	1
Den DF	499
Pr > F	0.1316
Sample Size = 500	

Rao-Scott Likelihood Ratio Test	
Likelihood Ratio Chi-Square	2.2750
Design Correction	1.0000
Rao-Scott Chi-Square	2.2750
DF	1
Pr > ChiSq	0.1315
F Value	2.2750
Num DF	1
Den DF	499
Pr > F	0.1321
Sample Size = 500	

Contingency tables for data from a complex survey. The only difference between using the SURVEYFREQ procedure to construct contingency tables and compute odds ratios for an SRS and doing so for a complex sample is that for the complex sample we include one of the following:

- the WEIGHT, STRATA, and CLUSTER statements to calculate linearization variances, or
- the WEIGHT and REPWEIGHTS statements to calculate variances using a replication method.

Example 10.5 of SDA. Code 10.2 shows how to use the SURVEYFREQ procedure to produce statistics for a two-factor contingency table when observations are from a stratified multistage sample—in this case, from the Survey of Youth in Custody data. The NOMCAR option is included in the SURVEYFREQ statement because some of the observations are missing values for at least one of the variables in the TABLES statement.

Code 10.2. Contingency table for complex survey data (`example1005.sas`).

```
PROC SURVEYFREQ DATA = syc NOMCAR;
   WEIGHT finalwt;
   STRATA stratum;
   CLUSTER psu;
   TABLES famtime*everviol / OR DEFF WCHISQ;
RUN;
```

Output 10.2(a) shows the entries of the contingency tables given in Example 10.5 of SDA. It also gives the estimated design effect (DEFF) for each estimated cell and margin percentage (see Section 7.3). You can calculate deffs for the marginal row and column percentages by including options ROW(DEFF) and COLUMN(DEFF) behind the slash in the TABLES statement.

Note that because of item nonresponse, the weighted frequencies underestimate the population totals. The missing values could be imputed if estimated population totals are desired. The estimated percentages are based on the cases with non-missing data.

Output 10.2(a). Contingency table for complex survey data (`example1005.sas`).

Table of famtime by everviol							
famtime	everviol	Frequency	Weighted Frequency	Std Err of Wgt Freq	Percent	Std Err of Percent	Design Effect
1	0	496	4838	497.11869	19.5878	1.4611	3.5061
	1	826	7946	446.44915	32.1713	1.5693	2.9196
	Total	1322	12784	737.51225	51.7592	1.5994	2.6503
2	0	484	4761	488.85666	19.2761	1.3559	3.0567
	1	782	7154	503.97531	28.9647	1.6131	3.2716
	Total	1266	11915	795.04232	48.2408	1.5994	2.6503
Total	0	980	9599	898.48685	38.8639	2.2890	5.7049
	1	1608	15100	783.56318	61.1361	2.2890	5.7049
	Total	2588	24699	1318	100.0000		
Frequency Missing = 33							

Output 10.2(b) gives the odds ratio and its confidence interval, from the OR option. The odds ratio is calculated as $(\hat{N}_{11}\hat{N}_{22})/(\hat{N}_{12}\hat{N}_{21})$, where $\hat{N}_{ij}$ is the estimated population total (weighted frequency) for the (i, j) cell. The column 1 relative risk is the ratio of (estimated probability of being in column 1 for units in row 1) to (estimated probability of being in column 1 for row 2), that is, $[\hat{N}_{11}/(\hat{N}_{11} + \hat{N}_{12})]/[\hat{N}_{21}/(\hat{N}_{21} + \hat{N}_{22})]$.

Output 10.2(b). Odds ratio for complex survey data (`example1005.sas`).

Odds Ratio and Relative Risks (Row1/Row2)			
Statistic	Estimate	95% Confidence Limits	
Odds Ratio	0.9149	0.7679	1.0900
Column 1 Relative Risk	0.9471	0.8510	1.0541
Column 2 Relative Risk	1.0352	0.9669	1.1084
Sample Size = 2588			

The confidence interval for the odds ratio includes 1, which indicates that there is no evidence of an association between the two factors. This is the same conclusion indicated by the p-value of 0.32 given by the Wald chi-square test requested by the WCHISQ option and shown in Output 10.2(c).

Output 10.2(c). Wald chi-square test statistic (`example1005.sas`).

Wald Chi-Square Test	
Chi-Square	0.9951
F Value	0.9951
Num DF	1
Den DF	835
Pr > F	0.3188
Sample Size = 2588	

10.2 Chi-Square Tests

The SURVEYFREQ procedure performs all of the chi-square tests and produces all of the measures of association discussed in SDA. Chi-square test statistics and p-values are requested as options following a slash (/) in the TABLES statement.

Table 10.1 lists some of the test statistics and measures of association produced. Code 10.2 requested the Wald chi-square test.

TABLE 10.1
Chi-square test statistics calculated by the SURVEYFREQ procedure.

Keyword	Statistic or Test
WALD	Wald test.
CHISQ	First-order Rao–Scott test, based on Pearson's chi-square test statistic (Rao and Scott, 1981, 1984).
CHISQ(SECONDORDER)	Second-order Rao–Scott test.
CHISQ(MODIFIED)	First-order Rao–Scott test, using the null hypothesis proportions to compute the correction to the test statistic.
LRCHISQ	First-order Rao–Scott likelihood ratio test.
LRCHISQ(SECONDORDER)	Second-order Rao–Scott likelihood ratio test.
LRCHISQ(MODIFIED)	First-order Rao–Scott likelihood ratio test, using the null hypothesis proportions to compute the correction to the test statistic.

Finite population corrections and chi-square tests. In general, I do not recommend using a finite population correction (fpc) when conducting a chi-square test. Often, the purpose of the test is to explore whether there is a general association between the factors in the superpopulation, not merely one in the finite population from which the data are drawn. Conducting the test without the fpc allows generalization to the superpopulation (under some superpopulation models) while still accounting for the clustering, stratification, and unequal weights in the sampling design. I usually omit the TOTAL= option in the SURVEYFREQ statement when carrying out hypothesis tests.

Example 10.6 of SDA. For this example, the variable *currviol* is defined as 1 if *crimtype* = 1 and 0 otherwise. This means that the "0" category of *currviol* consists of the persons with *crimtype* $\in \{2, 3, 4, 5\}$ as well as the 12 persons with missing values for *crimtype*, and can be thought of as the persons not known to have committed a violent offense. The analysis results are almost the same when the 12 missing values are excluded.

Code 10.3. Carry out a first-order Rao–Scott test (`example1006.sas`).

```
PROC SURVEYFREQ DATA = syc NOMCAR;
   WEIGHT finalwt;
   STRATA stratum;
   CLUSTER psu;
   TABLES currviol * ageclass / CHISQ DEFF;
RUN;
```

Output 10.3(a). Estimated percentages and design effects for table entries and margins (`example1006.sas`).

Table of curviol by ageclass							
curviol	ageclass	Frequency	Weighted Frequency	Std Err of Wgt Freq	Percent	Std Err of Percent	Design Effect
0	1	364	4247	854.74298	16.9798	2.8312	14.8978
	2	740	6542	582.36561	26.1554	1.7127	3.9791
	3	348	3190	224.80498	12.7539	1.2152	3.4771
	Total	1452	13979	1142	55.8892	2.5337	6.8223
1	1	238	2770	366.47247	11.0747	1.3269	4.6840
	2	516	4630	537.44003	18.5111	1.9301	6.4706
	3	415	3633	248.10343	14.5250	1.3478	3.8335
	Total	1169	11033	687.42539	44.1108	2.5337	6.8223
Total	1	602	7017	1067	28.0545	3.3395	14.4762
	2	1256	11172	916.85679	44.6666	2.6528	7.4598
	3	763	6823	354.42572	27.2789	2.2366	6.6068
	Total	2621	25012	1333	100.0000		

In Output 10.3(b), giving the first-order Rao–Scott test, the design correction is 2.4427 $= \hat{E}[X^2]/2$. The Rao–Scott statistic $X_F^2 = 13.92$ is obtained by dividing the Pearson chi-square statistic ($= 33.99$) in the first line of the output by the design correction. The F statistic is then obtained as $X_F^2/[(r-1)(c-1)]$.

Output 10.3(b). First-order Rao–Scott chi-square test statistic (`example1006.sas`).

Rao-Scott Chi-Square Test	
Pearson Chi-Square	33.9926
Design Correction	2.4427
Rao-Scott Chi-Square	13.9161
DF	2
Pr > ChiSq	0.0010
F Value	6.9580
Num DF	2
Den DF	1690
Pr > F	0.0010
Sample Size = 2621	

We can also run Code 10.3 with CHISQ(MODIFIED), which uses the null hypothesis proportions to compute the design correction (not shown). This gives slightly different statistics (although these differences would not affect conclusions). The modified design correction is 2.7388, which leads to $X_F^2(\text{mod}) = 33.99/2.7388 = 12.41$ and p-value 0.002.

Code 10.4 calculates the second-order Rao–Scott test statistics and p-values. The code is identical except for the test requested, which is now CHISQ(SECONDORDER). I recommend using the second-order test when there is sufficient information to calculate it; in general, p-values from the second-order test are closer to those that would be calculated from the exact distribution of the test statistic. Here, the second-order chi-square test statistic is $X_S^2 = 10.86$; the F statistic is 6.2.

Code 10.4. Carry out a second-order Rao–Scott test (`example1006.sas`).

```
PROC SURVEYFREQ DATA = syc NOMCAR;
   WEIGHT finalwt;
   STRATA stratum;
   CLUSTER psu;
   TABLES currviol * ageclass / CHISQ(SECONDORDER) DEFF;
RUN;
```

Output 10.4. Second-order Rao–Scott chi-square test statistic (`example1006.sas`).

Rao-Scott Chi-Square Test	
Pearson Chi-Square	33.9926
Design Correction	2.7388
First-Order Chi-Square	12.4116
Second-Order Chi-Square	10.8602
DF	1.75
Pr > ChiSq	0.0032
F Value	6.2058
Num DF	1.75
Den DF	1478.75
Pr > F	0.0033
Sample Size = 2621	

10.3 Loglinear Models

The SURVEYFREQ procedure does not, as of this writing, fit loglinear models for complex survey data. But one can use the CATMOD procedure to estimate coefficients for a loglinear model, and then use replication variance methods to calculate variances of the statistics. Let's first look at the CATMOD procedure for data from an SRS, where multinomial sampling can be assumed.

Example 10.8 of SDA. We use the data set *cable2*, with 500 records, that was created in Code 10.1. The CATMOD procedure will fit models to data in the form of the *cable1* data set, but the procedure with the data in the form of *cable2* can be extended for use with weights from a survey, as we shall see in Example 10.9.

Code 10.5 shows the CATMOD procedure code for independent data. Each observation is assumed to have weight 1. The MODEL statement is of the form:

```
MODEL response_effect = _RESPONSE_ / options;
```

In Code 10.5, the *response_ effect* is the table cross-classifying the variables *cable* and *computer*. The keyword _RESPONSE_ tells the procedure that you want to model the variation among the variables in the cross-classified table. You specify the model to be used in the LOGLIN statement. Here, it is desired to fit an additive model so the LOGLIN statement contains the variables *cable* and *computer*. The option in the MODEL statement requests a table of the predicted table probabilities under the independence model.

Code 10.5. Fit the loglinear model of independence to the cable data (`example1008.sas`).

```
PROC CATMOD DATA = cable2;
   MODEL cable*computer = _RESPONSE_ / PRED = PROB;
   LOGLIN cable computer;
RUN;
```

Check the log file to confirm that the iterative method used to calculate the estimates converged. The log file also says that the default method used for calculation is maximum likelihood, which is the preferred method.

```
NOTE: The default estimation method for this model is maximum-likelihood.
NOTE: Maximum likelihood computations converged.
```

Output 10.5(a) shows the parameter estimates from the independence model. The estimates in the table are the coefficients for *cable* = "n" and *computer* = "n". The value of the likelihood ratio chi-square statistic is 2.27, with p-value = 0.13, just as was found in the chi-square test from Output 10.1.

Output 10.5(a). Parameter estimates from the loglinear model of independence with the cable data (`example1008.sas`).

Maximum Likelihood Analysis of Variance			
Source	DF	Chi-Square	Pr > ChiSq
cable	1	25.53	<.0001
computer	1	14.64	0.0001
Likelihood Ratio	1	2.27	0.1315

Analysis of Maximum Likelihood Estimates					
Parameter		Estimate	Standard Error	Chi-Square	Pr > ChiSq
cable	n	-0.2321	0.0459	25.53	<.0001
computer	n	0.1737	0.0454	14.64	0.0001

Output 10.5(b) shows the predicted probabilities calculated under the loglinear model with independent factors.

Output 10.5(b). Predicted probabilities from the loglinear model of independence with the cable data (`example1008.sas`).

Maximum Likelihood Predicted Values for Probabilities						
		Observed		Predicted		
cable	computer	Probability	Standard Error	Probability	Standard Error	Residual
n	n	0.21	0.0182	0.2262	0.0153	-0.016
n	y	0.176	0.017	0.1598	0.0124	0.0162
y	n	0.376	0.0217	0.3598	0.0186	0.0162
y	y	0.238	0.019	0.2542	0.0163	-0.016

We can calculate the parameter estimates for the saturated model by including *cable*computer* in the LOGLIN statement, as in Code and Output 10.6.

Code 10.6. Fit a saturated loglinear model to the cable data (`example1008.sas`).

```
PROC CATMOD DATA = cable2;
   MODEL cable*computer = _RESPONSE_ / PRED = PROB;
   LOGLIN cable computer cable*computer;
RUN;
```

Output 10.6. Parameter estimates from saturated loglinear model with the cable data (`example1008.sas`).

Analysis of Maximum Likelihood Estimates					
Parameter		Estimate	Standard Error	Chi-Square	Pr > ChiSq
cable	n	-0.2211	0.0465	22.59	<.0001
computer	n	0.1585	0.0465	11.61	0.0007
cable*computer	n n	-0.0702	0.0465	2.28	0.1313

Example 10.9 of SDA. The CATMOD procedure can be used with survey weights, which will produce the correct point estimates for parameters but does not account for stratification or clustering in the survey design. However, you can use replicate weights to calculate standard errors of the parameter estimates. Let's do that with the saturated model for data from the Survey of Youth in Custody.

Step 1. Create the replicate weights. First, we need to create replicate weights that can be used to estimate variances. For the *syc* data, we could use the random group method, but that method has only 7 df for variance estimates. A full jackknife, deleting each person in turn in single-facility strata 6–16, gives correct variances but has a large number (861) of replicate weights, which greatly increases the size of the data set and slows calculations. Here, I created replicates using a grouped jackknife. Using "pseudo-psus" of the observations in facilities 6–16 reduces the number of replicates for which calculation is needed but still gives consistent estimates of the variances. I created variable *modpsu*, which equals *psu* for strata 1–5, and randomly groups residents into 7 pseudo-psus for each of the single facilities that serve as strata 6–16. Code 10.7 then creates 116 jackknife replicate weight variables.

Code 10.7. Construct jackknife weights for the data (`example1009.sas`).

```
PROC SURVEYMEANS DATA=syc VARMETHOD=JK(OUTWEIGHT=sycjkwts OUTJKCOEF=sycjkcoef);
  WEIGHT finalwt;
  STRATA stratum;
  CLUSTER modpsu;
  VAR age; /* Can use any variable for purpose of creating jackknife weights */
RUN;
```

If the data set you are using comes with replicate weights, you can skip Step 1.

Step 2. Find the parameter estimates for the full data set using the survey weights. Code 10.8 shows the CATMOD procedure code for the saturated model, using the survey weights in the WEIGHT statement. This code produces the parameter estimates in Output 10.8: the numbers in the Estimate column are correct for the survey, but the test statistics, standard errors, and p-values are wrong because they do not account for the survey design. This survey has high design effects because the facilities have a high degree of clustering with respect to ages served and severity of offenses.

The LOGLIN statement in Code 10.8 requests the saturated model (alternatively, you could type "*ageclass|everviol|famtime*" to include all interactions among the factors). The estimated probabilities in Example 10.9 of SDA are given by the predicted probabilities that are output to data set *predsat*. The ODS OUTPUT statement requests that the parameter estimates from the model be stored in data set *estsat*. I also request that the convergence status be output to a data set.

Code 10.8. Fit the saturated loglinear model to the Survey of Youth in Custody. This gives the point estimates, but the standard errors are incorrect (`example1009.sas`).

```
PROC CATMOD DATA = syc;
   WEIGHT finalwt;
   MODEL ageclass*everviol*famtime = _RESPONSE_ / PRED=PROB;
   LOGLIN ageclass everviol famtime ageclass*everviol ageclass*famtime everviol*
    famtime ageclass*everviol*famtime;
   ODS OUTPUT PREDICTEDPROBS = predsat ESTIMATES = estsat CONVERGENCESTATUS =
    convsat;
RUN;
```

Step 3. Find the parameter estimates using each set of replicate weights. This is most easily done using a macro such as the one given in Appendix B. It runs the CATMOD procedure in Code 10.8 with each *RepWt* variable in turn, and stores the estimates in a data set.

Step 4. Use Equation (9.1) to calculate the replication standard error. Output 10.8(b) shows the jackknife standard errors for the model parameters. The test statistics are calculated as $\hat{\theta}_i/\text{SE}(\hat{\theta}_i)$ for each estimated parameter, and the p-values are calculated using a t distribution. The p-values differ slightly from those in Table 10.5 of SDA, which were calculated using linearization variances in R software, but lead to similar conclusions.

Output 10.8(b). Jackknife standard errors for coefficients in saturated model (`example1009.sas`).

Obs	Parameter	ClassValue	estimatefull	jkstderr	tstat	pvalue
1	ageclas*everviol*famtime	1 0 1	0.00888	0.02966	0.29950	0.76510
2	ageclas*everviol*famtime	2 0 1	0.01613	0.02845	0.56707	0.57177
3	ageclass	1	-0.11486	0.11979	-0.95888	0.33963
4	ageclass	2	0.34409	0.07479	4.60095	0.00001
5	ageclass*everviol	1 0	0.13656	0.06228	2.19266	0.03035
6	ageclass*everviol	2 0	0.07237	0.03464	2.08903	0.03891
7	ageclass*famtime	1 1	0.05545	0.03792	1.46235	0.14637
8	ageclass*famtime	2 1	0.01281	0.02903	0.44122	0.65989
9	everviol	0	-0.24459	0.04763	-5.13479	0.00000
10	everviol*famtime	0 1	-0.03170	0.02361	-1.34243	0.18210
11	famtime	1	0.02423	0.03328	0.72797	0.46811

Additional ways to fit loglinear models in SAS software. A loglinear model is a special case of Poisson regression, and SAS Institute Inc. (2015) shows how to use the jackknife with the GENMOD procedure to obtain parameter estimates and standard errors that account for the survey design. Silva (2017) has developed a macro, %SURVEYGENMOD, that will incorporate the survey weights, stratification, and clustering when calculating coefficients for generalized linear models; see also Silva and Silva (2014).

If a model-based analysis is desired, you can use the GLIMMIX procedure to incorporate the clustering explicitly into the model. Section 5.3 described how the MIXED procedure can be used to perform model-based linear regression analyses with clustered data; the GLIMMIX procedure extends this capability to generalized linear models such as Poisson regression models (see Zhu, 2014; Diaz-Ramirez et al., 2020).

10.4 Summary, Tips, and Warnings

The SURVEYFREQ procedure estimates probabilities in cross-classified contingency tables using the weights. It also carries out tests for independence of factors that account for the survey design. Methods for adjusting the test statistics and p-values include the Wald test and the first- and second-order Rao–Scott tests.

The general form of the SURVEYFREQ procedure for analyzing two-way tables is as follows:

```
PROC SURVEYFREQ DATA=sample_dataset NOMCAR;
   WEIGHT weight_variable;             /* Always include a weight variable */
   STRATA stratification_variables; /* Describes the stratification of the psus */
   CLUSTER cluster_variable;           /* The cluster_variable identifies the psus */
   TABLES row_factor * column_factor / table_options;
```

If desired, the estimates can be calculated using replicate weights instead of specifying the design with STRATA and CLUSTER statements. Include the JKCOEFS= option in the REPWEIGHTS statement if the jackknife method is used.

```
PROC SURVEYFREQ DATA=sample_dataset VARMETHOD = variance_est_method (options);
   WEIGHT weight_variable;          /* Always include a weight variable */
   REPWEIGHTS RepWt_1 - RepWt_R; /* Lists the replicate weight variables */
   TABLES row_factor * column_factor / table_options;
```

Either formulation allows numerous options that can be listed behind the slash (/) in the TABLES statement. Table 10.2 lists some of the options that are useful for calculations in SDA. Additional options are described in the documentation for the SURVEYFREQ procedure (SAS Institute Inc., 2021).

Although SAS software does not, as of this writing, have a survey procedure for fitting loglinear models, these can be fit by using the non-survey procedure CATMOD along with replicate weights to calculate the variances of parameters. Various authors have also written macros that can be used to estimate loglinear model coefficients for survey data.

Tips and Warnings

- Always include a WEIGHT statement in the SURVEYFREQ procedure. If the survey design includes stratification and clustering, also include STRATA and CLUSTER statements (or include the information needed to calculate variances using a replication method). Check that the sum of the weights output by the procedure approximately equals the population size.

- To estimate contingency tables and test association of two factors within a domain, include the domain variable as the first variable cross-classified in the TABLES statement. Thus, to carry out a second-order Rao–Scott test for the independence of factors *row-factor* and *column-factor* separately for each level of variable *domain-variable*, write:

```
TABLES domain-variable * row-factor * column-factor / CHISQ(SECONDORDER);
```

- It is often desirable to omit the TOTAL= option when conducting a chi-square test on survey data. This results in variance estimates that are calculated without a finite population correction.

- When fitting a loglinear model, check that the estimates converged. These are computed using an iterative procedure; if the procedure does not converge, the estimates are not meaningful. Fienberg and Rinaldo (2007) discuss convergence issues in loglinear models.

TABLE 10.2

Some options for the TABLES statement in the SURVEYFREQ procedure. See Table 10.1 for options that can be used with the chi-square tests.

Option	Description
WALD	Wald test.
CHISQ	Rao–Scott tests based on Pearson's chi-square test statistic.
LRCHISQ	Rao–Scott likelihood ratio tests.
OR	Calculate odds ratios and relative risks.
CL	Calculate confidence intervals for estimated percentages. If estimated percentages are close to 0 or 100, use CL(TYPE=CLOPPERPEARSON) or CL(TYPE=WILSON).
CLWT	Calculate confidence intervals for estimated population totals (weighted frequencies) in each cell.
ROW	Calculate proportions as a fraction of each row total. To obtain confidence intervals and design effects, use ROW(CL DEFF).
COL	Calculate proportions as a fraction of each column total.
PLOTS=	Display selected plots. Options include a plot of odds ratios, a weighted frequency plot, and a mosaic plot.

11

Regression with Complex Survey Data

11.1 Straight Line Regression in an SRS

For many analyses carried out on a simple random sample (SRS), results from a model-based analysis in a procedure designed for independent and identically distributed data (such as the FREQ procedure) are the same as the results from the corresponding survey analysis procedure (such as the SURVEYFREQ procedure) used with weights set equal to 1. For regression, however, the standard errors for an SRS calculated in the REG or GLM procedures, which perform model-based regression analyses (see Section 4.6), differ from those in the SURVEYREG procedure. This is because, as explained in Section 11.2 of SDA, the standard errors for the SRS calculated using linearization account for the errors in estimating the population totals of both the x and y variables; the model-based standard error calculated in the GLM procedure is conditional on the values of x in the sample and calculated under the model assumptions. We can see the difference for the estimates calculated in Examples 11.2 and 11.4 of SDA using the two procedures.

Example 11.2 of SDA. Code and Output 11.1 show the commands and regression parameter estimates from the GLM procedure for the data in `anthsrs.csv`. This conducts a model-based analysis under assumptions (A1)–(A4) given in Section 11.1 of SDA. The CLPARM option in the MODEL statement requests confidence intervals for the regression parameters. The procedure also produces an ANOVA table and graphs of the data and the residuals (not shown here).

Code 11.1. Model-based regression for an SRS (GLM procedure; `example1102.sas`).

```
PROC GLM DATA = anthsrs;
  MODEL height = finger/ CLPARM;
RUN;
```

Output 11.1. Model-based regression for an SRS (GLM procedure; `example1102.sas`).

R-Square	Coeff Var	Root MSE	height Mean
0.487903	2.670214	1.749791	65.53000

Parameter	Estimate	Standard Error	t Value	Pr > |t|	95% Confidence Limits	
Intercept	30.31624803	2.56681185	11.81	<.0001	25.25445021	35.37804585
finger	3.04525031	0.22171730	13.73	<.0001	2.60801991	3.48248071

Example 11.4 of SDA. Now let's perform the regression analysis using the SURVEYREG procedure (Code and Output 11.2). The MODEL statement is identical to that used in

DOI: 10.1201/9781003160366-11

Code 11.1. The major difference for this SRS is the inclusion of the WEIGHT statement, where the variable wt is set equal to 3000/200 for each observation. Code 11.2 also produces a shaded plot of the data, analogous to that in Output 7.14 (not shown here).

Code 11.2. Design-based regression analysis for an SRS (SURVEYREG procedure; `example1104.sas`).

```
PROC SURVEYREG DATA = anthsrs;
   WEIGHT wt;
   MODEL height = finger/ CLPARM;
RUN;
```

Output 11.2. Design-based regression analysis for an SRS (SURVEYREG procedure; `example1104.sas`).

Fit Statistics	
R-Square	0.4879
Root MSE	1.7498
Denominator DF	199

Estimated Regression Coefficients						
Parameter	Estimate	Standard Error	t Value	Pr > |t|	95% Confidence Interval	
Intercept	30.3162480	2.55004716	11.89	<.0001	25.2876659	35.3448302
finger	3.0452503	0.22062698	13.80	<.0001	2.6101835	3.4803171

Note: The degrees of freedom for the t tests is 199.

The value of R^2 and the estimated slope and intercept are the same as in the analysis with the GLM procedure, but the standard errors of the regression coefficients, now calculated using linearization, are different. In this example, where the straight-line model fits the data well, the difference in the standard errors is small. In other examples the two sets of standard errors may exhibit wider disparities.

Degrees of freedom (df) for regression analyses. One other difference between the two analyses is the default df used for the confidence intervals. For the model-based analysis in Example 11.2, a t distribution with $n - 2 = 198$ df is used; for the analysis in the SURVEYREG procedure, a t distribution with $n - 1 = 199$ df is used.

The default df in the SURVEYREG procedure equals (number of psus minus number of strata) when linearization is used or when replication weights are constructed in the procedure from the design information (no REPWEIGHTS statement). When a REPWEIGHTS statement is included, the df equals the number of REPWEIGHTS variables. I use the default df for most analyses—most samples are large enough that the df value makes little difference. But I use the DF= option in the MODEL statement to change the default if: (1) the number of replicate weights supplied through the REPWEIGHTS statement far exceeds the number of sampled psus (this commonly occurs when the bootstrap is used), or (2) an analysis is being done on a domain that does not appear in some of the psus. Valliant and Rust (2010) discuss df considerations for complex surveys.

Example 11.6 of SDA. We can also calculate jackknife weights for the survey in the SURVEYREG procedure and use those to compute the standard errors. The SURVEYREG procedure in Code 11.3 calculates standard errors using the jackknife and also creates the jackknife weights in the data set specified by the OUTWEIGHTS option (these are 3000/199 = 15.0754 for the observations not deleted for the replicate and 0 for the observation that is deleted) and stores the jackknife coefficients in the data set specified by OUTJKCOEFS.

Code 11.3. Regression for SRS with jackknife (`example1106.sas`).

```
PROC SURVEYREG DATA = anthsrs VARMETHOD = JK (OUTWEIGHTS = anthsrsjkwt OUTJKCOEFS=
    anthsrsjkcoef);
   WEIGHT wt;
   MODEL height = finger/ CLPARM;
RUN;
```

The output for the jackknife is similar to that in Output 11.2 (where standard errors are calculated using linearization), but the jackknife standard errors are slightly larger. This is not a matter for concern; the two methods of variance estimation are asymptotically equivalent but often give slightly different numbers for real, finite-sized, data sets.

Output 11.3. Regression for SRS with jackknife (`example1106.sas`).

Fit Statistics	
R-Square	0.4879
Root MSE	1.7498
Denominator DF	199

Estimated Regression Coefficients

Parameter	Estimate	Standard Error	t Value	Pr > |t|	95% Confidence Interval	
Intercept	30.3162480	2.58045582	11.75	<.0001	25.2277013	35.4047948
finger	3.0452503	0.22329977	13.64	<.0001	2.6049129	3.4855878

Note: The degrees of freedom for the t tests is 199.

Finite population corrections for regression analyses. If desired, you can include a TOTAL= option in the SURVEYREG statement to calculate standard errors that incorporate a finite population correction (fpc). For Code 11.2 and 11.3, this will result in standard errors that are smaller by a factor of $\sqrt{1-200/3000}$. I usually omit the TOTAL= option when performing regression analyses, because I want to learn about the relationships among variables in a universal sense (including potential populations that are similar to the finite population), not just in the particular finite population that was studied. Ask yourself: If I were estimating regression relationships for data from a population census, would I want the standard errors of the coefficients to be zero (as they would be if a census were taken because there is no sampling variability)? If the answer is no, then omit the fpc.

11.2 Linear Regression with Complex Survey Data

We already used the SURVEYREG procedure in Chapter 4 to find regression estimates and in Chapter 7 to produce scatterplots. Now let's look at using it to calculate regression coefficients and provide other summary statistics and diagnostics for regression analyses with complex survey data having stratification, clustering, or unequal weights.

11.2.1 Straight Line Regression

Examples 11.3, 11.5, and 11.6 of SDA. We can calculate the straight-line regression relationship for the data in `anthuneq.csv` by including the WEIGHT variable in the SURVEYREG procedure (the STRATA and CLUSTER statements are not needed here because this design did not involve stratification or clustering). Code 11.4 performs the jackknife calculations in Example 11.6 of SDA. If analyzing data for which replicate weights have already been calculated, simply include the VARMETHOD option and the WEIGHT and REPWEIGHTS statements as in Code 9.17.

Code 11.4. Regression for an unequal-probability sample with jackknife variance estimation (`example1106.sas`).

```
PROC SURVEYREG DATA = anthuneq VARMETHOD=JK (OUTWEIGHTS = anthuneqjkwt OUTJKCOEFS=
    anthuneqjkcoef);
   WEIGHT wt;
   MODEL height = finger/ CLPARM;
RUN;
```

The code to obtain linearization standard errors displayed in Table 11.2 of SDA is similar and is given in `example1105.sas` on the website; for linearization estimates, simply substitute VARMETHOD=TAYLOR in the SURVEYREG procedure statement. The SURVEYREG procedure in Code 11.4 also produces a shaded plot of the data that includes the fitted regression line. Additional code provided in `example1105.sas` (not shown here) produces a bubble plot similar to that in Figure 11.5 of SDA.

For this example, the linearization standard errors, shown in Table 11.2 of SDA, are about 13% smaller than the jackknife standard errors given in Output 11.4. It is unusual for the two sets of standard errors to be this discrepant—remember, the jackknife and linearization variances are asymptotically equivalent, so for large samples, we expect the standard errors from jackknife to be approximately equal to those from linearization. In this data set, however, the observations with large y values have extremely large weights, and that gives them a lot of influence for determining the estimated regression line. In particular, the observation with *finger* = 12.2 and *height* = 71 is highly influential, and the estimated slope and intercept for the jackknife replicate deleting that observation differ substantially from the full-sample estimates. Most of the difference between the linearization and jackknife standard errors for this example can be attributed to that one influential observation.

Output 11.4. Regression for an unequal-probability sample with jackknife variance estimation (`example1106.sas`).

Fit Statistics	
R-Square	0.4400
Root MSE	1.8688
Denominator DF	199

Estimated Regression Coefficients						
Parameter	Estimate	Standard Error	t Value	Pr > |t|	95% Confidence Interval	
Intercept	30.1858583	7.62922789	3.96	0.0001	15.1413524	45.2303642
finger	3.0540995	0.67891576	4.50	<.0001	1.7153071	4.3928919

Note: The degrees of freedom for the t tests is 199.

11.2.2 Multiple Linear Regression

Now let's look at a regression analysis that includes more than one independent variable, for a survey that has unequal weights, stratification, and clustering.

Example 11.7 of SDA. Code 11.5 fits the regression model

$$bmxbmi = ridageyr + ridageyr^2$$

to the National Health and Nutrition Examination Survey (NHANES) data (`nhanes.csv`); *ridageyr* is the variable in the data that gives each person's age in years. This analysis fits the model to the entire data set; Code 11.9 will show an example of a regression model fit to the observations in the domain of adults.

Code 11.5. Regression of *bmxbmi* on *ridageyr*, *ridageyr*ridageyr* (`example1107.sas`).

```
PROC SURVEYREG DATA = nhanes NOMCAR;
  WEIGHT wtmec2yr;
  STRATA sdmvstra;
  CLUSTER sdmvpsu;
  MODEL bmxbmi = ridageyr ridageyr*ridageyr / SOLUTION CLPARM DEFF;
  OUTPUT OUT=quadout PRED=quadpred RESIDUAL=quadres;
RUN;
```

As in Code 7.4, we use the NOMCAR option to treat missing values as a separate domain for variance estimation purposes. The MODEL statement tells the procedure to fit a model with y variable *bmxbmi* and x variables *ridageyr* and *ridageyr*ridageyr*. It also asks explicitly for all parameter estimates (SOLUTION option) to be displayed with confidence intervals (CLPARM option) and estimated design effects (DEFF option). Residuals and predicted values are saved under variable names *quadres* and *quadpred* in the data set *quadout* for further analysis.

Output 11.5(a). Fit statistics and parameter estimates for regression of *bmxbmi* on *ridageyr*, *ridageyr*ridageyr* (`example1107.sas`).

Fit Statistics	
R-Square	0.2834
Root MSE	6.5007
Denominator DF	15

Let's first look at the fit statistics in Output 11.5(a). The SURVEYREG procedure calculates R^2 as

$$R^2 = 1 - \widehat{\text{SSW}}/\widehat{\text{SSTO}} = 1 - 12{,}881{,}829{,}731/17{,}975{,}165{,}897 = 0.2834,$$

where $\widehat{\text{SSW}} = \sum_{i\in\mathcal{S}} w_i(y_i - \hat{y}_i)^2 = 12{,}881{,}829{,}731$ estimates the within (residual) sum of squares from a fit of the regression model to the entire population, and $\widehat{\text{SSTO}} = \sum_{i\in\mathcal{S}} w_i(y_i - \hat{\bar{y}})^2 = 17{,}975{,}165{,}897$ estimates the total corrected sum of squares for the entire population.[1] The square root of the mean squared error (Root MSE in the fit statistics)

[1]The SURVEYREG procedure will calculate these sums of squares if you include the ANOVA option in the MODEL statement; I did not include the option in Code 11.5, however, because I typically use the population ANOVA table only for demonstrating how R^2 is calculated. Since R^2 is given in the fit statistics, the ANOVA table is usually not needed. However, the code in file `example1107.sas` includes the ANOVA option if you want to see it.

is calculated as $\sqrt{n\widehat{\text{SSW}}/\left[(n-p)\sum_{i\in\mathcal{S}} w_i\right]}$, where n is the number of observations and p is the number of parameters in the model (including the intercept). For most surveys, $n/(n-p) \approx 1$, and the Root MSE estimates the size of a "typical" residual if you were to fit the regression model to the entire population.

Output 11.5(b). Parameter estimates and tests for regression of *bmxbmi* on *ridageyr*, *ridageyr*ridageyr* (`example1107.sas`).

Tests of Model Effects			
Effect	Num DF	F Value	Pr > F
Model	2	746.83	<.0001
Intercept	1	4447.99	<.0001
ridageyr	1	1024.81	<.0001
ridageyr*ridageyr	1	618.88	<.0001

Note: The denominator degrees of freedom for the F tests is 15.

Estimated Regression Coefficients							
Parameter	Estimate	Standard Error	t Value	Pr > |t|	95% Confidence Interval		Design Effect
Intercept	15.2981488	0.22938064	66.69	<.0001	14.8092355	15.7870621	1.18
ridageyr	0.6031084	0.01883971	32.01	<.0001	0.5629525	0.6432643	2.48
ridageyr*ridageyr	-0.0057488	0.00023109	-24.88	<.0001	-0.0062413	-0.0052563	2.58

Output 11.5(b) shows the values of Wald F statistics for testing model effects, followed by the parameter estimates. For a linear hypothesis $H_0 : \mathbf{L}\boldsymbol{\beta} = 0$, the Wald F statistic is calculated as

$$F = \frac{(\mathbf{L}\hat{\mathbf{B}})^T \left[\mathbf{L}\,\hat{V}(\hat{\mathbf{B}})\,\mathbf{L}^T\right]^{-1} (\mathbf{L}\hat{\mathbf{B}})}{\text{rank}\left[\mathbf{L}\,\hat{V}(\hat{\mathbf{B}})\,\mathbf{L}^T\right]}$$

(a generalized inverse may be used when the inverse of $\hat{V}(\mathbf{L}\,\hat{\mathbf{B}})$ does not exist), and is compared to an F distribution with numerator df equal to the rank of $\mathbf{L}\,\hat{V}(\hat{\mathbf{B}})\,\mathbf{L}^T$. The denominator df for most surveys is (number of psus minus number of strata). In Output 11.5(b), the F statistic of 746.83 is for the null hypothesis that the coefficients of *ridageyr* and $ridageyr^2$ are both zero, and is compared to an $F_{2,15}$ distribution. It is common for tests about model parameters with large survey data sets to be highly significant (here, the p-value is much less than 0.0001) because the large sample size causes entries of $\hat{V}(\hat{\mathbf{B}})$ to be small.

The estimated regression coefficients are accompanied by their standard errors, and, with the DEFF option, their estimated design effects (deffs). The SURVEYREG procedure calculates deffs as (estimated variance of statistic under complex sampling design) / (estimated variance of statistic if an SRS of the same size were taken); see Buskirk (2011) for a detailed description of deff calculation. Note that the deffs for the regression coefficients are smaller than 6.34, the deff for the mean of body mass index (this deff can be calculated with an intercept-only model in the SURVEYREG procedure, using `MODEL BMXBMI = / DEFF;`). Deffs for estimated regression coefficients are often smaller than those for estimated population means or totals because the covariates in the regression model explain some of the cluster-to-cluster variability in the response variable (Lohr, 2014).

The SURVEYREG procedure will produce plots, accounting for the survey weights, for polynomial models involving a single explanatory variable. Output 11.5(c) displays a shaded plot with the fitted quadratic model.

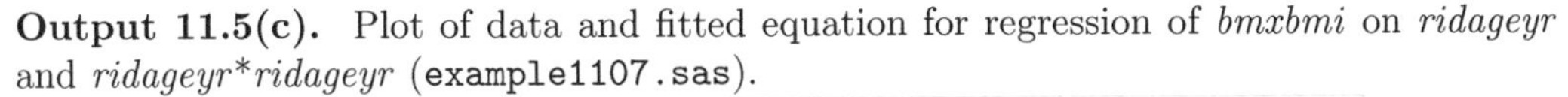

Output 11.5(c). Plot of data and fitted equation for regression of *bmxbmi* on *ridageyr* and *ridageyr***ridageyr* (`example1107.sas`).

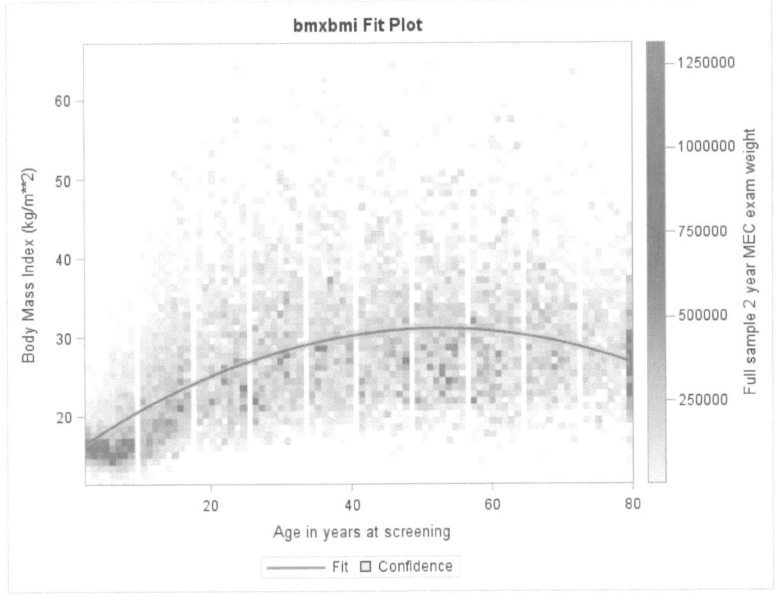

Output 11.5(d) shows the Data Summary produced by the SURVEYREG procedure. You should check this, as usual, to make sure that the sum of the weights is in the right neighborhood for the population size (it is, after accounting for missing data).

Output 11.5(d). Data summary for regression of *bmxbmi* on *ridageyr* and $ridageyr^2$ (`example1107.sas`).

Data Summary	
Number of Observations	9971
Number of Observations Used	8756
Number of Obs with Nonpositive Weights	427
Sum of Weights	304922183
Weighted Mean of bmxbmi	27.28165
Weighted Sum of bmxbmi	8318781460

Regression analyses and missing data. Output 11.5(d) also shows the number of observations used in the regression, 8756. The SURVEYREG procedure excludes an observation from the analysis if it is missing the value for the WEIGHT, STRATA, or CLUSTER variable, or if it is missing the value of the y-variable or any explanatory variable in the MODEL statement (see Section 8.1).

Here, almost all of the observations have values for *bmxbmi*. Other data sets or models, however, may exhibit more item nonresponse. Item nonresponse can distort estimates of regression relationships in the population. It is also possible that when a model has many explanatory variables, each with some item nonresponse, the missing data patterns can mesh in such a way that the model is fit on relatively few observations. Or the model might be fit on data from only a few of the psus. I recommend exploring the amount and patterns of missing data before performing analyses.

11.3 Using Regression to Compare Domain Means

The SURVEYREG procedure allows regression models to be fit with categorical explanatory variables as well as numeric explanatory variables. Simply list the categorical variables in a CLASS statement. Then, if the categorical variable *class_var1* has k levels, the following statements within the SURVEYREG procedure fit a regression model to $(k - 1)$ indicator variables that define the levels of the variable. The SOLUTION option requests explicit estimates for each level of the categorical variable.

```
CLASS class_var1;
MODEL y = class_var1 / SOLUTION;
```

You can include as many numeric and categorical variables as you like in the regression model. The following statements request a regression model predicting y from three numeric variables (*xvar1, xvar2, xvar3*) and two categorical variables along with their interaction.

```
CLASS class_var1 class_var2;
MODEL y = xvar1 xvar2 xvar3 class_var1 class_var2 class_var1*class_var2 / SOLUTION;
```

Example 11.8 of SDA. There are several ways to compare domain means in the SAS software survey analysis programs. I list three methods, labeled as methods D1, D2, and D3. All give the same estimates for differences of domain means, but one method may be easier to use for a particular analysis than another.

D1 Create indicator variables that explicitly define the domains, and use the indicator variables as covariates in the SURVEYREG procedure. This method gives the most flexibility and control for defining domains and contrasts, but also requires the most coding when there are multiple domains.

Code 11.6 fits the model predicting BMI from x, where x (defined in the DATA step to equal 1 if the person is female and 0 if the person is male) is treated as a numeric variable. The regression coefficient corresponding to x is the estimated difference between the mean BMI of females and the mean BMI of males. From Output 11.6, this value is 0.6896696 with a 95% CI of [0.1501024, 1.2292369], as shown in Table 11.4 of SDA.

Code 11.6. Perform regression analysis on the indicator variable x (`example1108.sas`).

```
DATA nhanes;
  SET datalib.nhanes;
  IF riagendr = 1 THEN x = 0; /* x=0 is male*/
  IF riagendr = 2 THEN x = 1; /* x=1 is female */

PROC SURVEYREG DATA = nhanes NOMCAR;
  WEIGHT wtmec2yr;
  STRATA sdmvstra;
  CLUSTER sdmvpsu;
  MODEL bmxbmi = x / CLPARM;
RUN;
```

Output 11.6. Perform regression analysis on the indicator variable x (`example1108.sas`).

Estimated Regression Coefficients						
Parameter	Estimate	Standard Error	t Value	Pr > |t|	95% Confidence Interval	
Intercept	26.9276308	0.20702536	130.07	<.0001	26.4863667	27.3688949
x	0.6896696	0.25314568	2.72	0.0157	0.1501024	1.2292369

Note: The degrees of freedom for the t tests is 15.

To apply this approach to a domain with k categories, you need to define $k-1$ indicator variables, where $xvar\_j = 1$ if the observation is in category j and 0 otherwise, for $j = 1, \ldots, k-1$. That can be cumbersome, and I recommend using (D2) for comparing multiple domain means.

D2 Declare the domain variable as a CLASS variable, and use that as the explanatory variable in the MODEL statement of the SURVEYREG procedure. Use the LSMEANS statement with the DIFF option to calculate CIs for the pairwise domain differences, as shown in Code 11.7. If comparing more than two domains, you can adjust the confidence intervals for multiple comparisons by including ADJUST=BON (for Bonferroni adjustments), ADJUST=TUKEY (for Tukey-type adjustments), or other multiple comparison method as an option following the slash in the LSMEANS statement. Hsu (1996) has a thorough treatment of the theory of multiple comparisons; Westfall et al. (2011) discuss how to implement multiple comparison methods in SAS software.

Code 11.7. Perform regression analysis on the class variable x (`example1108.sas`).

```
PROC SURVEYREG DATA = nhanes NOMCAR;
  CLASS x;
  WEIGHT wtmec2yr;
  STRATA sdmvstra;
  CLUSTER sdmvpsu;
  MODEL bmxbmi = x / NOINT SOLUTION CLPARM;
  LSMEANS x / DIFF;
RUN;
```

The NOINT option in the MODEL statement tells the procedure to fit a model without an intercept term, so that the parameter estimates are the estimated domain means, as seen in Output 11.7(a).

Output 11.7(a). Least squares means from regression on the class variable x (`example1108.sas`).

Estimated Regression Coefficients						
Parameter	Estimate	Standard Error	t Value	Pr > |t|	95% Confidence Interval	
x 0	26.9276308	0.20702536	130.07	<.0001	26.4863667	27.3688949
x 1	27.6173004	0.26005588	106.20	<.0001	27.0630044	28.1715964

Note: The degrees of freedom for the t tests is 15.

Output 11.7(b) shows the estimates for the two domain means (here, the same as the regression coefficients because the NOINT option was used) and their difference from the LSMEANS statement. This output gives the male minus female difference, so the signs of the difference and the confidence limits are reversed from those in Output 11.6.

If desired, you can change the reference category in the CLASS statement, as will be seen in Code 11.11.

Output 11.7(b). Least squares means from regression on the class variable x (`example1108.sas`).

x Least Squares Means								
x	Estimate	Standard Error	DF	t Value	Pr > |t|	Alpha	Lower	Upper
0	26.9276	0.2070	15	130.07	<.0001	0.05	26.4864	27.3689
1	27.6173	0.2601	15	106.20	<.0001	0.05	27.0630	28.1716

Differences of x Least Squares Means									
x	_x	Estimate	Standard Error	DF	t Value	Pr > |t|	Alpha	Lower	Upper
0	1	-0.6897	0.2531	15	-2.72	0.0157	0.05	-1.2292	-0.1501

D3 Use the CLDIFF option of the DOMAIN statement in the SURVEYMEANS procedure to request confidence intervals for the pairwise domain differences. You can request multiple comparisons using the Bonferroni method, if desired, by including the ADJUST=BON option in the DOMAIN statement.

Code 11.8. Find differences of domain means using the SURVEYMEANS procedure (`example1108.sas`).

```
PROC SURVEYMEANS DATA = nhanes NOMCAR MEAN CLM;
  WEIGHT wtmec2yr;
  STRATA sdmvstra;
  CLUSTER sdmvpsu;
  DOMAIN x / CLDIFF;
  VAR bmxbmi;
RUN;
```

Output 11.8 again gives the male minus female difference, so the signs of the difference and the confidence limits are reversed from those in Output 11.6.

Output 11.8. Find differences of domain means using the SURVEYMEANS procedure (`example1108.sas`).

The SURVEYMEANS Procedure

Statistics for x Domains						
x	Variable	Label	Mean	Std Error of Mean	95% CL for Mean	
0	bmxbmi	Body Mass Index (kg/m**2)	26.927631	0.207015	26.4863898	27.3688718
1	bmxbmi	Body Mass Index (kg/m**2)	27.617300	0.260042	27.0630334	28.1715673

Differences of bmxbmi (Body Mass Index (kg/m**2)) Means for x Domains								
x	-x	Diff Estimate	Std Error	DF	t Value	Pr > |t|	95% Confidence Limits	
0	1	-0.689670	0.253132	15	-2.72	0.0157	-1.2292086	-0.1501306

Comparing subdomain means within domains. The estimates in Output 11.6, 11.7, and 11.8 are for persons of all ages. What if you want to compare the mean value of BMI for adult men with the mean BMI for adult women, that is, to compare men and women within the domain of adults? Again, there are several ways to do this.

1. Define a domain variable (or set of indicator variables) to have three domains: adult men, adult women, and everyone else. Then use the SURVEYMEANS or SURVEYREG procedure to estimate the domain means, and compare the first two of those.

2. Define variable *adult* to equal 1 if the person is an adult and 0 otherwise. Then use the SURVEYREG procedure to estimate the means for men and women, and the difference of the means, for each level of *adult*. Code 11.9 declares *adult* to be the DOMAIN variable. Then three models predicting *bmxbmi* from x are fit: one with all the data, one for the domain with *adult* $= 0$, and one for the domain with *adult* $= 1$.

   Code 11.9. Perform regression on the class variable x separately for the domains of adults and nonadults (`example1108.sas`).

```
PROC SURVEYREG DATA = nhanes NOMCAR;
  CLASS x;
  WEIGHT wtmec2yr;
  STRATA sdmvstra;
  CLUSTER sdmvpsu;
  DOMAIN adult;
  MODEL bmxbmi = x / SOLUTION CLPARM;
  LSMEANS x / DIFF;
RUN;
```

   Output 11.9 shows the output from the LSMEANS statement for the domain with *adult*$= 1$. This gives the estimated mean BMI for adult women as 29.6443, the estimated mean BMI for adult men as 29.1116, and the estimated difference in means (men − women) as −0.5327.

   Output 11.9. Least squares means for domain with *adult* $= 1$ (`example1108.sas`).

x Least Squares Means

x	Estimate	Standard Error	DF	t Value	Pr > |t|	Alpha	Lower	Upper
0	29.1116	0.2619	15	111.17	<.0001	0.05	28.5534	29.6697
1	29.6443	0.2886	15	102.70	<.0001	0.05	29.0290	30.2595

Differences of x Least Squares Means

x	_x	Estimate	Standard Error	DF	t Value	Pr > |t|	Alpha	Lower	Upper
0	1	-0.5327	0.2204	15	-2.42	0.0289	0.05	-1.0025	-0.06288

3. Use replicate weights to estimate variances. Then you can perform one of the analyses in (D1) to (D3) on the subset of the data that is in the domain of interest, as discussed in Section 9.5. Code 9.8 created BRR replicate weights in data set *nhanesbrr*. Code 11.10 replaces the STRATA and CLUSTER statements with REPWEIGHTS. The output is similar to that in Output 11.9.

Code 11.10. Perform regression analysis on the class variable x on the adults in the data and use replicate weights (`example1108.sas`).

```
PROC SURVEYREG DATA = nhanesbrr VARMETHOD = BRR;
  WHERE adult = 1;  /* Perform the analysis on the subset of the data with adult
      = 1 */
  CLASS x;
  WEIGHT wtmec2yr;
  REPWEIGHTS repwt_1-repwt_16;
  MODEL bmxbmi = x/ SOLUTION CLPARM;
  LSMEANS x / DIFF;
RUN;
```

Example 11.9 of SDA. Domain comparison is similar with k domains. Code 11.11 shows a comparison of BMI for adults in five race/ethnicity groups measured in NHANES. First, we define the response categories and the domain variable in the data set. After that, the code is similar to that in (D2). The option REF= in the CLASS statement tells the procedure to use the first category of *raceeth*, when ordered alphabetically, as the reference category.[2]

Code 11.11. Perform regression on the class variable *raceeth* separately for the domains of adults and nonadults (`example1109.sas`).

```
DATA nhanes;
  SET datalib.nhanes;
  IF ridageyr >= 20 THEN adult = 1;
  ELSE adult = 0;
  IF ridreth3 IN (1,2) THEN raceeth = "Hispanic";
  ELSE IF ridreth3 IN (3) THEN raceeth = "White";
  ELSE IF ridreth3 IN (4) THEN raceeth = "Black";
  ELSE IF ridreth3 IN (6) THEN raceeth = "Asian";
  ELSE IF ridreth3 IN (7) THEN raceeth = "Other";

PROC SURVEYREG DATA = nhanes NOMCAR;
  CLASS raceeth / REF = first;
  WEIGHT wtmec2yr;
  STRATA sdmvstra;
  CLUSTER sdmvpsu;
  DOMAIN adult;
  MODEL bmxbmi = raceeth / SOLUTION CLPARM;
  LSMEANS raceeth / DIFF ADJUST=TUKEY;
RUN;
```

Code 11.11 produces several pieces of output for each domain. First, the MODEL statement gives the fit statistics, provides a Wald F statistic and p-value for the null hypothesis that all means are equal, and produces estimates of the regression parameters. These are shown in Output 11.11(a) for the domain with *adult* $= 1$. For this model, $R^2 = 0.033$. Even though the F statistic for testing equality of all group means is 131.57, indicating significant differences among at least some of the groups, race and ethnicity explain only a small part of the variability in BMI in the population.

[2]You can also fit the model with the NOINT option, as in Code 11.7, if you want the regression coefficients to be the 5 group means. However, the NOINT option also changes the values of the F statistic and the coefficient of determination R^2, so here I displayed the group means with the LSMEANS statement.

Output 11.11(a). Fit statistics and F test for comparing race/ethnicity groups among adults (`example1109.sas`).

Fit Statistics	
R-Square	0.03264
Root MSE	6.8609
Denominator DF	15

Tests of Model Effects			
Effect	Num DF	F Value	Pr > F
Model	4	131.57	<.0001
Intercept	1	29561.3	<.0001
raceeth	4	131.57	<.0001

Note: The denominator degrees of freedom for the F tests is 15.

Output 11.11(b) shows the regression coefficients from Table 11.5 of SDA. Note that the coefficient for the reference category, Asian, is 0. The other regression coefficients estimate the difference between the mean for the category listed and the reference category, and you can calculate each estimated domain mean from the intercept and regression coefficient for the domain.

Output 11.11(b). Parameter estimates for comparing race/ethnicity groups among adults (`example1109.sas`).

Estimated Regression Coefficients						
Parameter	Estimate	Standard Error	t Value	Pr > |t|	95% Confidence Interval	
Intercept	24.9706788	0.14116854	176.89	<.0001	24.6697852	25.2715724
raceeth Black	5.6205312	0.33808373	16.62	<.0001	4.8999227	6.3411396
raceeth Hispanic	5.6265680	0.33683525	16.70	<.0001	4.9086207	6.3445154
raceeth Other	5.4714137	0.53378751	10.25	<.0001	4.3336726	6.6091549
raceeth White	4.2598721	0.23723360	17.96	<.0001	3.7542207	4.7655235
raceeth Asian	0.0000000	0.00000000	.	.	0.0000000	0.0000000

However, if you want to compare the domain means, the LSMEANS statement is more convenient. It calculates estimates of the mean of each domain and also calculates pairwise differences of those means.

The output from the LSMEANS statement for the domain with *adult*$=1$ is displayed in Output 11.11(c). It shows the estimated mean for each group and gives confidence intervals for the pairwise differences. The ADJUST=TUKEY option asks that these confidence intervals be adjusted for multiple comparisons using the Tukey-Kramer method (Hsu, 1996).

Output 11.11(c). Least squares means for comparing race/ethnicity groups among adults (`example1109.sas`).

raceeth Least Squares Means								
raceeth	Estimate	Standard Error	DF	t Value	Pr > |t|	Alpha	Lower	Upper
Asian	24.9707	0.1412	15	176.89	<.0001	0.05	24.6698	25.2716
Black	30.5912	0.3366	15	90.88	<.0001	0.05	29.8737	31.3087
Hispanic	30.5972	0.3027	15	101.07	<.0001	0.05	29.9520	31.2425
Other	30.4421	0.4731	15	64.34	<.0001	0.05	29.4336	31.4506
White	29.2306	0.2595	15	112.64	<.0001	0.05	28.6774	29.7837

Differences of raceeth Least Squares Means Adjustment for Multiple Comparisons: Tukey-Kramer												
raceeth	_raceeth	Estimate	Standard Error	DF	t Value	Pr > |t|	Adj P	Alpha	Lower	Upper	Adj Lower	Adj Upper
Asian	Black	-5.6205	0.3381	15	-16.62	<.0001	<.0001	0.05	-6.3411	-4.8999	-6.6645	-4.5766
Asian	Hispanic	-5.6266	0.3368	15	-16.70	<.0001	<.0001	0.05	-6.3445	-4.9086	-6.6667	-4.5864
Asian	Other	-5.4714	0.5338	15	-10.25	<.0001	<.0001	0.05	-6.6092	-4.3337	-7.1197	-3.8231
Asian	White	-4.2599	0.2372	15	-17.96	<.0001	<.0001	0.05	-4.7655	-3.7542	-4.9924	-3.5273
Black	Hispanic	-0.00604	0.3635	15	-0.02	0.9870	1.0000	0.05	-0.7807	0.7687	-1.1284	1.1163
Black	Other	0.1491	0.5471	15	0.27	0.7889	0.9986	0.05	-1.0169	1.3151	-1.5402	1.8384
Black	White	1.3607	0.3277	15	4.15	0.0009	0.0065	0.05	0.6623	2.0590	0.3489	2.3724
Hispanic	Other	0.1552	0.5775	15	0.27	0.7918	0.9987	0.05	-1.0757	1.3860	-1.6281	1.9384
Hispanic	White	1.3667	0.3458	15	3.95	0.0013	0.0096	0.05	0.6296	2.1038	0.2988	2.4346
Other	White	1.2115	0.5479	15	2.21	0.0430	0.2281	0.05	0.04364	2.3794	-0.4805	2.9035

Diffplots for domain comparisons. The SURVEYREG procedure also produces a bivariate plot of the group means (Hsu and Peruggia, 1994) that is useful for assessing differences among the groups. The plot in Output 11.11(d), sometimes called a diffplot, shows light horizontal and vertical lines that display the position of each group mean relative to the vertical and horizontal axes. Each diagonal line going from northwest to southeast is centered at the intersection of the means for two of the groups, and its length is the width of the confidence interval for the difference of that pair of means multiplied by $\sqrt{2}$ (by the Pythagorean theorem, because the lines are on the diagonal). The dashed lines that intersect the reference line (diagonal line from southwest to northeast) indicate pairs of means that are not significantly different. The solid lines that do not intersect the reference line represent pairs of means that are significantly different. This plot is a graphical representation of the confidence intervals displayed numerically in Output 11.11(c).

In Output 11.11(d), the estimated mean BMI for Asian Americans is 24.97 kg/m^2, and that is represented by the leftmost vertical line and the bottommost horizontal line. The mean for white Americans is 29.23 kg/m^2. The diagonal line centered at (24.97, 29.23) shows the confidence interval for the difference in mean BMI for white and Asian Americans; the statistical significance of that difference can be seen on the graph because the confidence interval does not intersect with the southwest-to-northeast diagonal reference line.

Output 11.11(d). Diffplot for comparing BMI among race/ethnicity groups (`example1109.sas`).

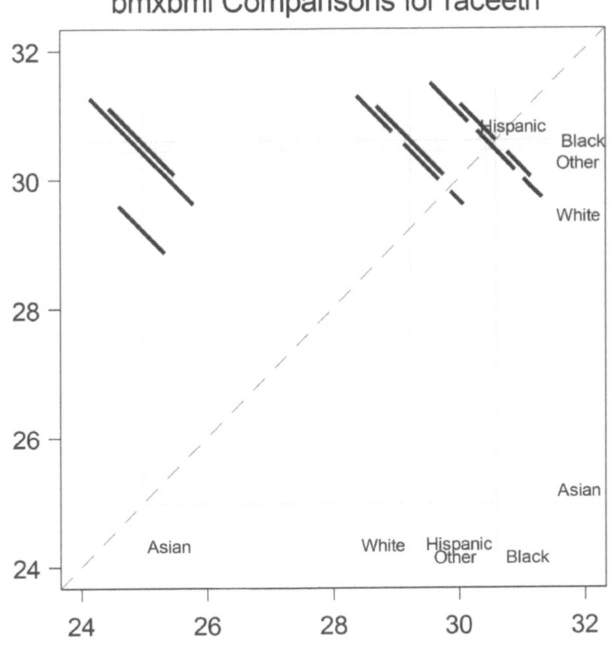

11.4 Logistic Regression

The SURVEYLOGISTIC procedure, which performs logistic regression, has syntax similar to the SURVEYREG procedure. The main difference is in the MODEL statement, since the dependent variable is categorical in a logistic regression. For the SURVEYLOGISTIC procedure, the MODEL statement has the form:

```
CLASS class_var1 class_var2;
MODEL y (EVENT = 'level') = xvar1 xvar2 xvar3 class_var1 class_var2;
```

The option (EVENT = 'level') tells the procedure to treat the value of y corresponding to 'level' as the value for which the probability is modeled. Thus, typing (EVENT = '1') tells the procedure to predict $P(Y = 1)$. The software has defaults for which level is modeled if you omit the EVENT option, but I find it easier to avoid confusion if I define the predicted level explicitly through the EVENT option.

11.4.1 Logistic Regression with a Simple Random Sample

Example 11.11 of SDA. Data set *cable2* was created in Code 10.1. We want to predict the probability that the household has a computer from the variable *cable*. Both of these are categorical, so the DATA step in Code 11.12 defines variable x to equal 1 if the household subscribes to cable and 0 otherwise. The MODEL statement asks the procedure to predict the probability that *computer* = 'y' from x (alternatively, one could use *cable* as the explanatory variable after declaring it to be categorical in a CLASS statement).

Code 11.12. Logistic regression for SRS (`example1111.sas`).

```
DATA cable2;
   SET cable2;
   IF cable = 'y' THEN x = 1;
   ELSE x = 0;

PROC SURVEYLOGISTIC DATA = cable2;
   WEIGHT wt;
   MODEL computer (EVENT = 'y') = x / CLPARM CLODDS;
RUN;
```

The output from the SURVEYLOGISTIC procedure includes information about the model, the variance estimation method, the sum of the weights, the number of observations used in the model, the model fit statistics, and estimates of the regression coefficients and odds ratios along with 95% confidence intervals (requested by the CLPARM and CLODDS options in the MODEL statement). This section displays the portions of the output used in Example 11.11 of SDA.

Output 11.12(a) shows the response profile and convergence status. The response profile tells you how many observations are in each response category and confirms that the procedure is predicting the probability that *computer* = 'y.' Checking this helps you interpret the signs of the logistic regression coefficients correctly. In this example, a significant positive coefficient for x would mean that larger values of x are associated with a higher predicted probability of having a computer.

The convergence status, which is also given in the log, says that the iterative procedure used to calculate the parameter estimates converged for this data set and model. You should always check the convergence status; if the procedure does not converge, you need to find out why and consider an alternative model. Allison (2008) reviews common causes of nonconvergence in logistic regression and tells how to fix it.

Output 11.12(a). Response profile and convergence status (`example1111.sas`).

Response Profile			
Ordered Value	computer	Total Frequency	Total Weight
1	n	293	293.00000
2	y	207	207.00000

Probability modeled is computer='y'.

Model Convergence Status
Convergence criterion (GCONV=1E-8) satisfied.

Output 11.12(b) shows the parameter estimates and odds ratio estimate for the model. The options CLPARM and CLODDS request confidence intervals for the estimates. The standard errors for the parameters are calculated using the Wald method discussed in Chapter 10. The odds ratio is calculated as $\exp(\hat{B}_1) = \exp(-0.2807) = 0.755$ and the confidence limits for the odds ratio are $0.524 = \exp(-0.6469)$ and $1.089 = \exp(0.0856)$.

For this example, the logistic regression coefficient for the log odds ratio x is not significantly different from 0, and consequently the 95% confidence interval for the odds ratio includes 1. There is no evidence that the variables *computer* and *cable* are associated.

Output 11.12(b). Regression coefficients and odds ratio estimate for logistic regression (`example1111.sas`).

Analysis of Maximum Likelihood Estimates

Parameter	Estimate	Standard Error	t Value	Pr > |t|	95% Confidence Limits	
Intercept	-0.1766	0.1448	-1.22	0.2232	-0.4611	0.1079
x	-0.2807	0.1864	-1.51	0.1328	-0.6469	0.0856

NOTE: The degrees of freedom for the t tests is 499.

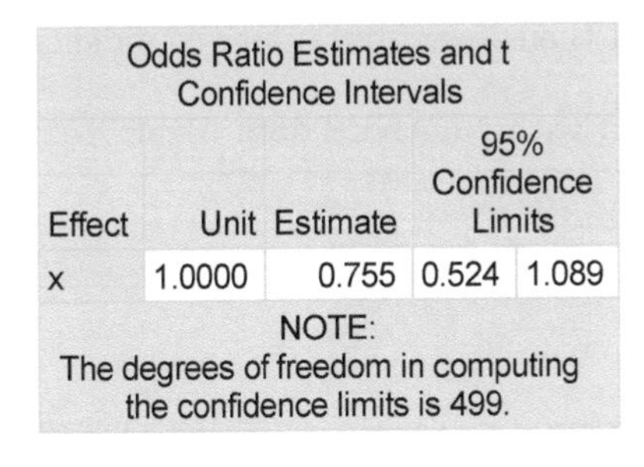

Odds Ratio Estimates and t Confidence Intervals

Effect	Unit	Estimate	95% Confidence Limits	
x	1.0000	0.755	0.524	1.089

NOTE:
The degrees of freedom in computing the confidence limits is 499.

The SURVEYLOGISTIC procedure also produces three types of test statistics for the null hypothesis that all parameters (other than the intercept) equal zero: a likelihood ratio test using the Rao–Scott adjustment, score test (see Rao et al., 1998), and Wald test. These are shown in Output 11.12(c). For this simple model fit to an SRS, with one binary covariate, the likelihood ratio and Wald test statistics equal G^2 from Example 10.1 of SDA.

Output 11.12(c). Test statistics for logistic regression (`example1111.sas`).

Testing Global Null Hypothesis: BETA=0

Test	F Value	Num DF	Den DF	Pr > F
Likelihood Ratio	2.27	1	499	0.1321
Score	2.26	1	499	0.1335
Wald	2.27	1	499	0.1328

11.4.2 Logistic Regression with a Complex Survey

You can specify the same variance estimation methods for the SURVEYLOGISTIC procedure as for the SURVEYREG or SURVEYMEANS procedure. This section demonstrates logistic regression using the default linearization variances. Replicate weights can also be used with the same syntax as for the SURVEYMEANS procedure illustrated in Section 9.4.

Example 11.12 of SDA. A logistic regression model is fit predicting the probability that BMI > 30 from gender (variable *female*) and waist circumference (variable *bmxwaist*) for the domain of adults in the NHANES data.

Before running the commands in Code 11.13, I define the binary response variable *bmi30* to equal 1 if BMI > 30 and 0 if BMI <= 30 (see Code 7.3) and define variable *female* to equal 1 if the person is a female and 0 if the person is a male. Here, *female* is treated as a numeric variable; it could also be analyzed as a categorical variable if desired by declaring it as such in the CLASS statement. The numeric variable *bmxwaist* is treated as a continuous covariate.

Code 11.13 has the same basic form as used for the analysis of the SRS in Code 11.12; the major change is adding the STRATA and CLUSTER statements that define the survey design, as well as the NOMCAR option so that missing data are treated as a domain. I include the DOMAIN statement because I want the regression to be limited to the domain with *adult* = 1.

The MODEL statement says to use an additive model to predict the probability. The model fit in this example, which does not include an interaction term, has the same rate of increase for the logit of the predicted probability for males and females, even though the two genders may have different base probabilities. I request option DF=design so that 15 df are used for the F tests shown in Output 11.13(d). The OUTPUT statement saves the predicted probabilities from the model and their lower and upper confidence limits.

Code 11.13. Logistic regression with NHANES data (`example1112.sas`).

```
PROC SURVEYLOGISTIC DATA = nhanes NOMCAR;
  WEIGHT wtmec2yr;
  STRATA sdmvstra;
  CLUSTER sdmvpsu;
  DOMAIN adult;
  MODEL bmi30 (EVENT = '1') = bmxwaist female / CLPARM CLODDS DF=design;
  OUTPUT OUT = lrpred PRED = predval LOWER = lcl UPPER = ucl;
RUN;
```

Again, the procedure produces lots of output, for the sample as a whole and for each domain. Here, I show the parts of the output that are relevant for the analysis in SDA. Output 11.13(a) shows the Domain Summary for the domain with *adult* = 1. This shows the sum of weights for the domain, which is approximately equal to the number of U.S. adults. The Model Convergence Status (not shown) states that the convergence criterion is satisfied, so we can proceed to look at the rest of the output.

Output 11.13(a). Domain summary for domain of adults (`example1112.sas`).

Domain Summary	
Number of Observations	9971
Number of Obs with Nonpositive Weights	427
Number of Observations in Domain	5474
Number of Observations not in Domain	4070
Sum of Weights in Domain	234507386

Output 11.13(b) and (c) show the parameter estimates and odds ratios for the model. Each odds ratio is calculated as $\exp(\hat{B}_j)$ for the corresponding regression coefficient. Thus the odds ratio for *bmxwaist* equals $\exp(0.2809) = 1.324$ and is interpreted as follows. Suppose that person 1 has a waist circumference that is 1 cm larger than the waist circumference of person 2, and the two persons have the same values for all of the other covariates in the model (in this example, they have the same gender). Then the model predicts the odds that person 1 has BMI > 30 to be 1.324 times as large as the odds that person 2 has BMI > 30.

Output 11.13(b). Parameter estimates for logistic regression with the NHANES data (`example1112.sas`).

Analysis of Maximum Likelihood Estimates						
Parameter	Estimate	Standard Error	t Value	Pr > |t|	95% Confidence Limits	
Intercept	-29.9557	1.1921	-25.13	<.0001	-32.4967	-27.4148
bmxwaist	0.2809	0.0115	24.43	<.0001	0.2564	0.3054
female	1.5786	0.1666	9.48	<.0001	1.2236	1.9337
NOTE: The degrees of freedom for the t tests is 15.						

Output 11.13(c). Odds ratios for logistic regression in NHANES (`example1112.sas`).

Odds Ratio Estimates and t Confidence Intervals				
Effect	Unit	Estimate	95% Confidence Limits	
bmxwaist	1.0000	1.324	1.292	1.357
female	1.0000	4.848	3.399	6.915
NOTE: The degrees of freedom in computing the confidence limits is 15.				

Output 11.13(d) gives the three tests that SAS software calculates for the null hypothesis that the effects of *bmxwaist* and *female* are both 0. By default, the SURVEYLOGISTIC procedure uses the second-order Rao–Scott correction for the likelihood ratio test if there are three or more parameters in the model (including the intercept). The Wald test statistic is also given, along with a score test statistic (Rao et al., 1998). All of the tests have p-value less than 0.0001, giving strong evidence against the null hypothesis.

Output 11.13(d). Tests for global null hypothesis that all parameters other than intercept are zero (`example1112.sas`).

Testing Global Null Hypothesis: BETA=0				
Test	F Value	Num DF	Den DF	Pr > F
Likelihood Ratio	1836.11	1.8030	27.0452	<.0001
Score	169.12	2	15	<.0001
Wald	312.20	2	15	<.0001
NOTE: Second-order Rao-Scott design correction 0.1093 applied to the Likelihood Ratio test.				

The OUTPUT statement in Code 11.13 saves the predicted values and their lower and upper confidence limits for the three models fit by the SURVEYLOGISTIC statement (the model for all the data, and the two domain models). Code 11.14 shows how to use that output to graph the predicted probability that BMI $>$ 30 for men (solid line) and women (dashed line).

Code 11.14. Plot the predicted probabilities from the logistic regression model (`example1112.sas`).

```
PROC SORT DATA = lrpred;
  BY female bmxwaist;

PROC SGPLOT DATA = lrpred;
  WHERE adult = 1 and domain = 'adult=1';
  SERIES X = bmxwaist Y = predval / GROUP = female;
  XAXIS MIN = 50 MAX = 175;
RUN;
```

Output 11.14. Predicted probability that BMI > 30, separately for men and women (`example1112.sas`).

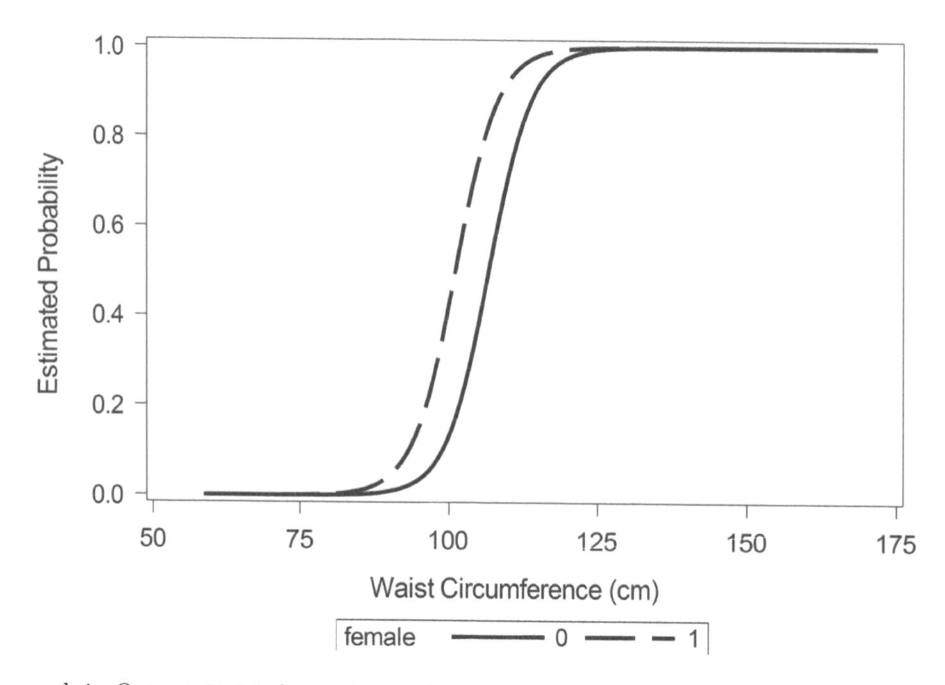

The graph in Output 11.14 shows the predicted values from the model. This does not show the original data, however, and a scatterplot of a binary variable y versus the explanatory variables usually does not provide much information about the relationship because it usually displays an indistinguishable mass of observations at $y = 1$ and another mass at $y = 0$. A more helpful option is to construct graphs showing the distribution of continuous covariates for each level of the response variable (here, *bmi30*). If there is a single continuous covariate, you may want to construct a histogram of the values of the covariate at each value of y.

Code 11.15 uses the SGPLOT procedure to draw side-by-side boxplots of waist circumference for the four domains formed by the cross-classification of gender (variable *female*) and the response variable *bmi30*. These boxplots could also be drawn using the SURVEYMEANS procedure with the DOMAIN statement. This shows the distribution of the continuous covariate *bmxwaist* for each level of the binary covariate *female* and each level of the response variable *bmi30*.

Code 11.15. Boxplots, using weights, of waist circumference for the four gender/BMI groups (`example1112.sas`).

```
DATA nhanes;
  SET nhanes;
  IF female = 0 AND bmi30 = 0 THEN group = "M BMI<=30";
  ELSE IF female = 1 AND bmi30 = 0 THEN group = "F BMI<=30";
  ELSE IF female = 0 AND bmi30 = 1 THEN group = "M BMI>30";
  ELSE IF female = 1 AND bmi30 = 1 THEN group = "F BMI>30";

PROC SGPLOT DATA = nhanes;
  WHERE adult = 1 AND bmxbmi ne .;
  HBOX bmxwaist / WEIGHT= wtmec2yr CATEGORY= group WHISKERPCT= 0 MEANATTRS= (SYMBOL
    = Plus);
  XAXIS LABEL = "Waist Circumference (cm)" MIN = 50 MAX = 175;
  YAXIS label = "Gender/BMI Category";
RUN;
```

Output 11.15 shows the difference in the distribution of waist circumference, separately for each gender, for the persons with high and low values of BMI. It is clear from the graph that waist circumference is a strong predictor of *bmi30*.

Output 11.15. Boxplots, using weights, of waist circumference for the four gender/BMI groups (`example1112.sas`).

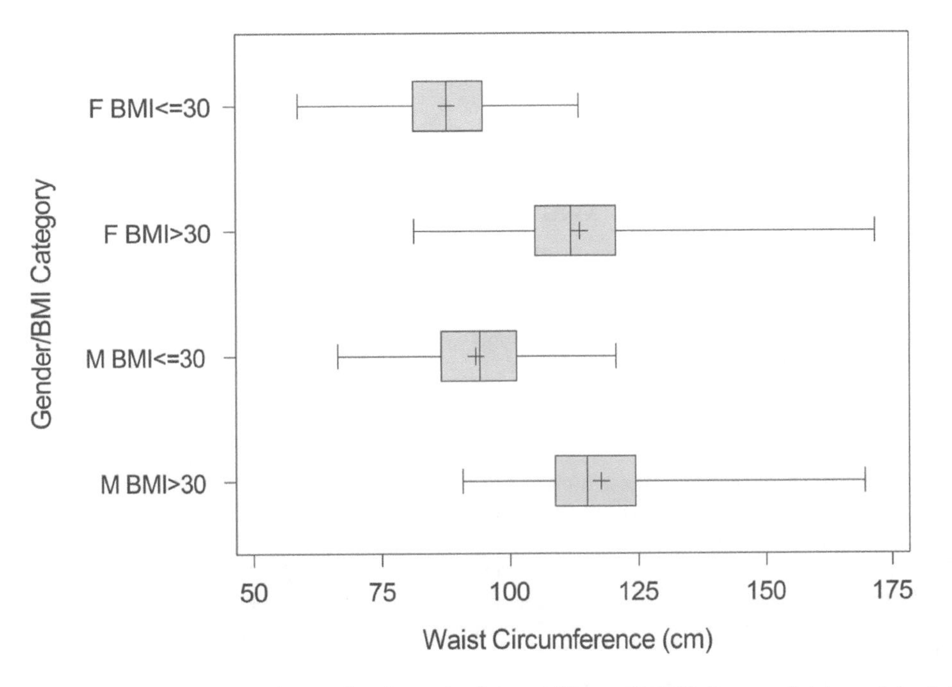

Many of the diagnostics and graphs described by Allison (2012) for model-based logistic regression can also be applied to survey data.

11.5 Additional Resources and Code for Exercises

Balanced sampling: Exercise 11.37 of SDA. There are several methods for selecting a balanced sample. The rejective method described in the exercise involves generating repeated samples, and then selecting a sample at random from those that meet the balancing constraints within a predetermined tolerance.

The rejective method can require a great deal of computation, however. If the population and sample are large, it may not be feasible to generate repeated samples and then reject those that fail to meet the balancing criteria. Chauvet and Tillé (2007) provide computationally efficient macros for selecting balanced samples using SAS software. See also Chauvet and Tillé (2006) and Tillé and Wilhelm (2017).

Mixed models for survey data. SAS software has several procedures for fitting mixed models. Stroup et al. (2018) provide a guide to using the MIXED and GLIMMIX procedures, which fit linear and logistic (and other) models to clustered data. Although those procedures do not allow for a full design-based approach for fitting mixed models, the GLIMMIX procedure can be used with survey weights. Zhu (2014) and Diaz-Ramirez et al. (2020) show how to use the GLIMMIX procedure with survey data.

Exercise 11.31 of SDA. The SURVEYREG procedure will create the matrix $\left(\sum_{j \in \mathcal{S}} w_j \mathbf{x}_j \mathbf{x}_j^T\right)^{-1}$ needed for calculating the leverage if you include the INVERSE option in the MODEL statement. You can also save the standard errors of the predicted values, along with the residuals and predicted values, by using the OUTPUT statement.

11.6 Summary, Tips, and Warnings

The SURVEYREG and SURVEYLOGISTIC procedures perform linear and logistic regression analyses with numeric and categorical explanatory variables.

The main statements used for a typical analysis with the SURVEYREG procedure, when linearization variances are desired, are given below. Many other statements and options are available for the SURVEYREG procedure, and these are described in detail in the SAS software documentation.

```
PROC SURVEYREG DATA=sample_dataset NOMCAR;
   CLASS class_var1 ... class_varp; /* Optional; include the CLASS statement if
     analyzing any categorical variables */
   WEIGHT weight_variable;          /* Always include a weight variable */
   STRATA stratification_variables; /* Describes the stratification of the psus  */
   CLUSTER cluster_variable;        /* The cluster_variable identifies the psus  */
   DOMAIN domain_variable;          /* Optional; include the DOMAIN statement if
     separate regression equations are desired for domains */
   MODEL y =  var1 var2 ... vark class_var1 ... class_varp / SOLUTION CLPARM;
   /* var1 var2 ... vark are the numeric explanatory variables and class_var1 ...
     class_varp are the categorical explanatory variables */
```

The optional NOMCAR keyword in the SURVEYREG statement tells the procedure to treat missing data as a separate domain for variance estimation; the procedure will simply ignore the NOMCAR option if the data set has no missing values, so it does no harm to include it. The SOLUTION and CLPARM options in the MODEL statement tell the procedure to create a table of the parameter estimates with confidence intervals for each.

If you wish to compare group means, where the groups are defined by the levels of the categorical variable *class_var1*, include an LSMEANS statement. The DIFF option asks the procedure to calculate pairwise differences for the group means, and the CL option requests confidence intervals for the group means and differences. If you would like to use a multiple comparison method, you can also include an ADJUST= option in the LSMEANS statement.

```
LSMEANS class_var1 / DIFF CL ADJUST=adjust_method;
```

The main statements used for an analysis with the SURVEYREG procedure, when the survey data includes replicate weights, are:

```
PROC SURVEYREG DATA=sample_dataset PLOTS=ALL VARMETHOD=varmethod_keyword;
   CLASS class_var1 ... class_varp; /* Optional; include the CLASS statement if
    analyzing any categorical variables */
   WEIGHT weight_variable;          /* Always include a weight variable */
   REPWEIGHTS repwt_1 - repwt_k;    /* Always include the replicate weight
    variables */
   DOMAIN domain_variable;          /* Optional; include the DOMAIN statement if
    separate regression equations are desired for domains */
   MODEL y =  var1 var2 ... vark class_var1 ... class_varp / SOLUTION CLPARM;
```

The key differences between the syntax for a replicate weight analysis and the syntax for a linearization-variance analysis are:

1. For replication variance estimation, specify which variance estimation method is used in the VARMETHOD= option. The choices for `varmethod_keyword` are JACKKNIFE, BOOTSTRAP, and BRR. If you omit the VARMETHOD= option but include the REPWEIGHTS statement, the procedure will use the jackknife method.
2. The REPWEIGHTS statement gives the names of the variables containing the k replicate weights. If using the jackknife method, also include the JKCOEFS= option, which supplies the values of the jackknife coefficients. If using replicate weights, do not include the STRATA or CLUSTER statement.

The SURVEYLOGISTIC procedure follows a similar format, with just a couple of differences. For the SURVEYLOGISTIC procedure, the MODEL statement has the form:

```
MODEL y (EVENT = 'level') =  var1 var2 ... vark class_var1 ... class_varp / CLPARM
    CLODDS;
```

The CLPARM and CLODDS options request confidence intervals for the regression parameters and odds ratios, respectively. Additional options are available that specify convergence criteria, degrees of freedom, and a generalized R^2 statistic.

Tips and Warnings

- If your data set has item missing data, check the Data Summary from the SURVEYREG or SURVEYLOGISTIC procedure to see how many observations were used in the analysis. If many observations were excluded from the model because they were missing y or one of the x variables, you might want to consider an alternative model for the data.

- If separate regression models are desired for domains, and if linearization variances are calculated (that is, replication is not used to estimate variances), include the DOMAIN statement in the SURVEYREG or SURVEYLOGISTIC procedure. If replication methods are used for variance estimation, you can fit separate regression models either by including the DOMAIN statement or by performing separate regression analyses using the replicate weights that have been created on the full data set.
- If performing logistic regression through the SURVEYLOGISTIC procedure, always check whether the model converged. The procedure will give you a warning in the output and the log that "The maximum likelihood estimate may not exist," but in some cases it produces estimates anyway, with the warning "Validity of the model fit is questionable." Heed this warning, and investigate why the estimates did not converge.
- Use `(EVENT= 'prediction.level')` in the MODEL statement of the SURVEYLOGISTIC procedure to specify the level of the variable for which the probability is predicted. If you omit this, the output will tell you which level is predicted, but it is better to control this in the program so that the coefficients correspond to the outcome you want to predict.

12

Additional Topics for Survey Data Analysis

12.1 Two-Phase Sampling

This section considers a simple two-phase design in which phase I is a simple random sample (SRS) and phase II is a stratified random sample. The survey analysis procedures in SAS software do not yet have special commands for two-phase samples, although they will provide summary statistics that can be plugged into variance formulas, as shown below. Another option is to use the macro *jk2phase*, given in Appendix B, to construct jackknife replicate weights using the method of Kim et al. (2006).

Examples 12.1 and 12.4 of SDA. The SURVEYMEANS procedure will calculate the two-phase point estimate $\hat{\bar{y}}^{(2)}_{\text{str}}$ in Equation (12.5) of SDA using the weight in variable *finalwt*, but the standard errors will not reflect the variability due to the phase I sampling. We can use the MEANS procedure to calculate the stratum summary statistics needed for the variance formula, as shown in Code 12.1.

Code 12.1. Calculate summary statistics for variance estimation (`example1201.sas`).

```
PROC MEANS DATA = vietnam N NMISS MEAN VAR;
   CLASS apc;
   VAR vietnam;
RUN;
```

Output 12.1. Summary statistics for each stratum (`example1201.sas`).

The MEANS Procedure

Analysis Variable : vietnam					
apc	N Obs	N	N Miss	Mean	Variance
No	804	72	732	0.1527778	0.1312598
NotAvail	505	505	0	0.4178218	0.2437294
Yes	755	67	688	0.7313433	0.1994573

Output 12.1 gives the values in Table 12.2 of SDA, which are then plugged into Equation (12.7) of SDA to calculate the variance of $\hat{\bar{y}}^{(2)}_{\text{str}}$.

Jackknife variance estimation. Alternatively, we can use the jackknife to obtain a with-replacement approximation to the variance. Code 12.2 uses the macro *jk2phase* to create phase II jackknife weights with the procedure in Section 12.4 of SDA.

DOI: 10.1201/9781003160366-12

Code 12.2. Create two-phase jackknife weights (`example1201.sas`).

```
%INCLUDE "C:\MyFilePath\JK2phase.sas"; /* Load the macro */

%jk2phase(indata=vietnam, p1jkwt=vp1jkwt, p2jkwt=vp2jkwt, jkcoefs=vjkcoefs, numreps
    =2064, strat=apc, nh=p1apcsize, mh=p2apcsize, p1weight=phase1wt, p2sample=
    p2sample);
```

The arguments of the macro are defined in Section B.2. For this example, the macro creates a set of phase I replicate weights for the *vietnam* data in data set *vp1jkwt*, a set of phase II replicate weights in data set *vp2jkwt*, and the jackknife coefficients in data set *vjkcoefs* (the same coefficients are used with the phase I replicate weights and the phase II replicate weights). The remaining arguments tell the macro which variables in the data set represent the stratum membership, phase I and phase II sample sizes in each stratum, phase I weights, and membership in the phase II sample.

For two-phase sampling with the delete-one jackknife, n (= phase I sample size) replicate weights are created for each phase. This can lead to slow computations when the phase I sample is large (but it is still usually faster than doing variance calculations by hand).

Now the weights can be used to calculate the standard error for the estimated percentage of veterans who served in Vietnam, in Code 12.3. Once the jackknife weights are created, they are used exactly as other jackknife weights.

Code 12.3. Use two-phase jackknife weights to calculate standard errors (`example1201.sas`).

```
PROC SURVEYMEANS DATA = vp2jkwt VARMETHOD = JACKKNIFE;
   WEIGHT finalwt;
   REPWEIGHTS repwt_1 - repwt_2064 / JKCOEF = vjkcoefs;
   VAR vietnam;
RUN;
```

Output 12.3. Use two-phase jackknife weights to calculate standard errors (`example1201.sas`).

The SURVEYMEANS Procedure

Data Summary	
Number of Observations	644
Sum of Weights	10320

Variance Estimation	
Method	Jackknife
Replicate Weights	VP2JKWT
Number of Replicates	2064

Statistics					
Variable	N	Mean	Std Error of Mean	95% CL for Mean	
vietnam	644	0.429262	0.027257	0.37580861	0.48271611

12.2 Estimating the Size of a Population

12.2.1 Ratio Estimation of Population Size

As discussed in Section 13.1 of SDA, the simple two-sample capture-recapture estimate can be calculated using ratio estimation.

Example 13.1 of SDA. In this example, a sample of 200 fish was caught, marked, then released. The marked and released fish were allowed to mix with the other fish in the lake, and then a second, independent, sample of 100 fish was caught. Twenty of the fish in the second sample were marked, having also been caught in the first sample. Code 12.4 shows how to use the RATIO statement in the SURVEYMEANS procedure for Example 13.1.

Code 12.4. Two-sample capture-recapture estimation using ratio estimation (`example1301.sas`).

```
%LET n1 = 200;
%LET n2 = 100;
%LET m = 20;

DATA fish;
   RETAIN y n1 wt;
   wt = 1;
   n1 = &n1;
   DO i = 1 TO &n2;
      IF i <= &m THEN x = 1;
      ELSE x = 0;
      OUTPUT;
   END;

PROC SURVEYMEANS DATA=fish PLOTS=NONE RATIO VAR CLM;
   WEIGHT wt;
   VAR n1 x;
   RATIO n1 / x;
```

Output 12.4. Two-sample capture-recapture estimation using ratio estimation (`example1301.sas`).

Ratio Analysis						
Numerator	Denominator	Ratio	Std Error	Var	95% CL for Ratio	
n1	x	1000.000000	201.007563	40404	601.157386	1398.84261

The DATA step creates the data set for the second sample of size n_2. I used macro variables so you can see how the code corresponds to the textbook formulas. Define variable *n1* to equal n_1 for all values. Then the ratio $\sum_{i=1}^{n_2}(n1)_i / \sum_{i=1}^{n_2} x_i = n_1 n_2/m$ and the RATIO statement will give standard errors and confidence intervals using the t distribution approximation. Note that the WEIGHT variable is set to a relative weight of 1.

The normal-approximation-based confidence interval works well with large samples but, as argued in Section 13.1 of SDA, can have inaccurate coverage probability when n_2 is small.

The file example1301.sas contains code (not shown here) for calculating a confidence interval by inverting a chi-square test using the procedure of Cormack (1992).

Code 12.5 shows how to create percentile confidence intervals using the bootstrap. There are several ways of implementing the bootstrap in SAS software. Chapter 9 discussed how to create bootstrap replicate weights in the SURVEYMEANS procedure. Here, I use an alternative method and employ the SURVEYSELECT procedure to draw bootstrap samples from the *fish* data. I take 2000 resamples with replacement from the original sample, calculate Chapman's estimate

$$\tilde{N} = \frac{(n_1 + 1)(n_2 + 1)}{m + 1} - 1$$

from each resample, and then use the 2000 estimates from the bootstrap resamples to estimate the sampling distribution of $\tilde{N}$. I use Chapman's estimate here because it is guaranteed to be finite for every bootstrap resample, whereas $\hat{N} = n_1 n_2 / m$ might be infinite if a resample contains no marked fish.

Code 12.5. Two-sample capture-recapture estimation with bootstrap (example1301.sas).

```
/* Use the SURVEYSELECT procedure to generate bootstrap samples */
%LET nboot = 2000;
PROC SURVEYSELECT DATA=fish OUT = fishbootsamp METHOD = URS SAMPSIZE = &n2 OUTHITS
     REPS=&nboot SEED=382073;
   TITLE 'Use the SURVEYSELECT procedure to draw bootstrap samples';

PROC SORT DATA = fishbootsamp;
   BY replicate;

PROC MEANS DATA = fishbootsamp NOPRINT SUM;
   BY replicate;
   VAR x;
   OUTPUT OUT = Outstats SUM = sum_x;

DATA Outstats;
   SET Outstats;
   Ntilde = (&n1 + 1)*(&n2 + 1)/(sum_x + 1) - 1;

PROC SGPLOT DATA = Outstats;
   HISTOGRAM Ntilde / binwidth = 100;
   TITLE 'Histogram of bootstrapped estimates of N';
   XAXIS LABEL = "Estimate of N";

PROC UNIVARIATE DATA=OutStats noprint;
   VAR Ntilde;
   OUTPUT OUT=Pctl pctlpre = CI95_
     pctlpts =2.5  97.5
     pctlname=Lower Upper;
   TITLE 'Calculate bootstrap percentile-based confidence interval';

PROC PRINT DATA = Pctl noobs;
RUN;
```

The bootstrap takes repeated samples of size n_2 with replacement from the original sample of size n_2, so METHOD=URS is used (see Table 6.1) to select each of the REPS=&nboot samples (you can use any value for REPS= that you like; here, I set `&nboot=2000`). The OUTHITS statement says to repeat each unit in the output data *fishbootsamp* as many times as it is sampled. After running the SURVEYSELECT procedure, the log tells you that the output data set has 200,000 (= 2000 n_2) records.

```
NOTE: The data set WORK.FISHBOOTSAMP has 200000 observations
      and 6 variables.
```

The SURVEYSELECT procedure creates a variable named *Replicate* that indexes the bootstrap samples. The next step is to calculate the desired statistic(s) from each replicate sample. The data set *OutStats* contains 2000 values of the estimated population size *Ntilde*. The SGPLOT procedure draws a histogram of these 2000 values, shown in Output 12.5(a), and the UNIVARIATE procedure calculates the 2.5th and 97.5th percentiles to form the bootstrap confidence interval in Output 12.5(b).

Output 12.5(a). Histogram of population size estimates from bootstrap samples (`example1301.sas`).

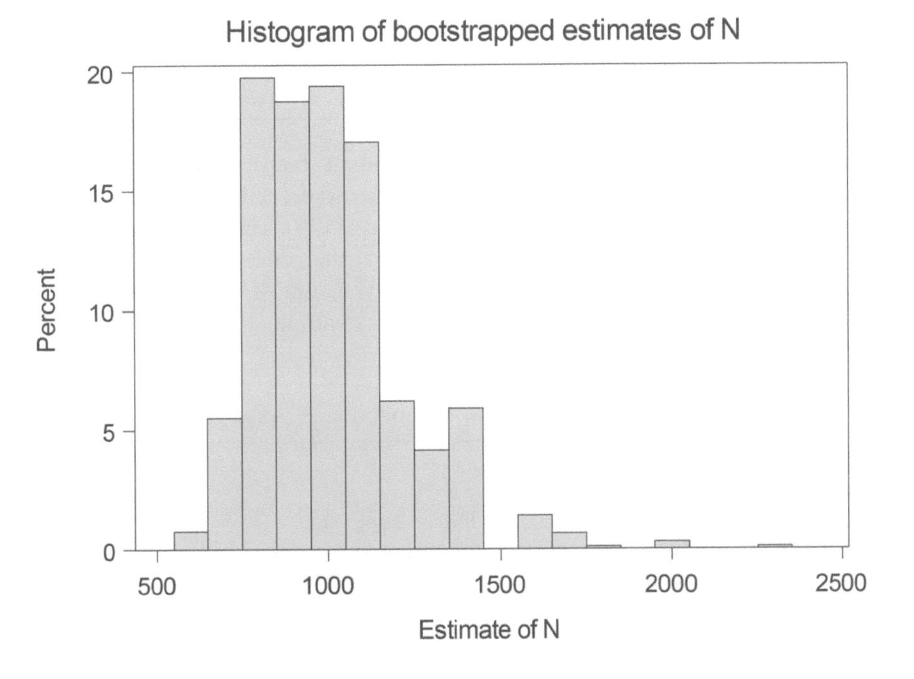

Output 12.5(b). Two-sample capture-recapture with bootstrap (`example1301.sas`).

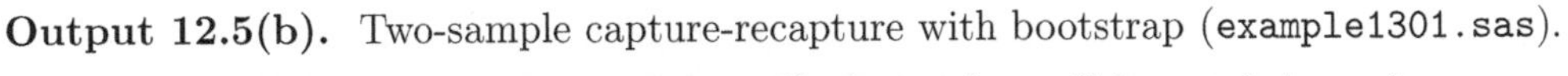

CI95_Lower	CI95_Upper
699.034	1560.62

12.2.2 Loglinear Models with Multiple Lists

Section 10.3 used the CATMOD procedure to fit loglinear models to data from a complex survey. When loglinear models are used in multiple-recapture estimation, it is often assumed that the lists are simple random samples. This section uses the CATMOD procedure to fit loglinear models thought to describe the dependence structure of the lists.

Example 13.3 of SDA. Each population list in the data set *opium* is denoted by a separate variable (here, variables *elist*, *dlist*, and *tlist*), and the variable *count* contains the number of units in the set of lists with value 1. In Code 12.6, there are 712 observations on the *tlist* but neither of the others, 69 observations in the *dlist* but neither of the others, and so on, with 6 observations on all three lists. The first line of the data shows a missing count for the units not on any of the lists.

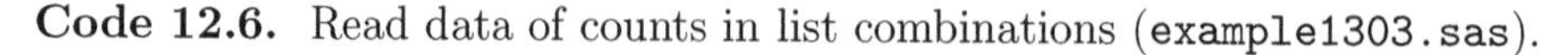

Code 12.6. Read data of counts in list combinations (`example1303.sas`).

```
DATA opium;
   INPUT elist dlist tlist count;
   DATALINES;
0 0 0 .
0 0 1  712
0 1 0   69
1 0 0 1728
0 1 1    8
1 0 1  314
1 1 0   27
1 1 1    6
;
```

Code 12.7 fits a loglinear model with three independent factors. In Section 10.3, the WEIGHT statement in the CATMOD procedure was used to describe the sampling weights. Here, the WEIGHT variable *count* contains the number of observations with the combination of factor levels in data set *opium*. In both instances, the procedure acts as though the data set contains the number of observations given by the sum of the weights. The *opium* data actually has that number of observations, so for this example, the CATMOD procedure gives correct inferences.

Code 12.7. Three-factor loglinear model with independence (`example1303.sas`).

```
PROC CATMOD DATA = opium;
   WEIGHT count;
   MODEL elist*dlist*tlist = _RESPONSE_ / PRED = freq ML;
   LOGLIN elist dlist tlist;
   TITLE 'Model of independent factors';
RUN;
```

The MODEL statement contains the names of the lists, joined by asterisks before the "=" sign. The keyword *_RESPONSE_* tells the procedure to fit a loglinear model. The option PRED requests the predicted values of the cell counts under the model, and ML asks for maximum likelihood computations.

The LOGLIN statement defines which loglinear model effects are to be fit. Code 12.7, listing the three factors, fits the model

$$\ln m_{ijk} = \mu + \alpha_i + \beta_j + \gamma_k$$

described in Section 13.2 of SDA.

Output 12.7 gives the maximum likelihood estimates of the parameters μ, α_i, β_j, and γ_k, as well as the value of $G^2 = 1.80$.

Output 12.7. Three-factor loglinear model with independence (`example1303.sas`).

Maximum Likelihood Analysis of Variance			
Source	DF	Chi-Square	Pr > ChiSq
elist	1	169.50	<.0001
dlist	1	1548.05	<.0001
tlist	1	865.96	<.0001
Likelihood Ratio	3	1.80	0.6154

Analysis of Maximum Likelihood Estimates					
Parameter		Estimate	Standard Error	Chi-Square	Pr > ChiSq
elist	0	0.4147	0.0319	169.50	<.0001
dlist	0	2.0562	0.0523	1548.05	<.0001
tlist	0	0.8585	0.0292	865.96	<.0001

The other models are fit similarly. To include an interaction term, include the two factors joined by an asterisk. Thus, Code 12.8 fits the loglinear model where *dlist* and *tlist* are conditionally independent given *elist*.

Code 12.8. Three-factor loglinear model with *elist* and *dlist* dependent, *elist* and *tlist* dependent (`example1303.sas`).

```
PROC CATMOD DATA = opium;
   WEIGHT count;
   MODEL elist*dlist*tlist = _RESPONSE_ / PRED = freq ML;
   LOGLIN elist dlist tlist elist*dlist elist*tlist;
RUN;
```

12.3 Small Area Estimation

Many researchers have implemented algorithms for computing small area estimates. Rao and Molina (2015), Pratesi (2016), and Tzavidis et al. (2018) describe some of the macros and packages that have been developed to fit small area models with SAS and R software. Some small area models can be fit using standard procedures in SAS software; Mukhopadhyay and McDowell (2011) show how to use the MIXED, IML, and MCMC procedures to fit area-level and unit-level models.

Statistics Canada has developed a comprehensive system that performs small area analyses in SAS software (Hidiroglou et al., 2019). The system allows for the calculation of unit-level and area-level models that, if desired, incorporate the survey design. It also allows the user to ensure that estimated totals for small areas sum to estimated design-based totals for larger areas (for which the survey has a sufficiently large sample size to give reliable estimates by itself) and to perform diagnostics on the model to explore its adequacy.

12.4 Evolving Capabilities of SAS Software

The capabilities of the survey design and analysis procedures in SAS software continue to expand as new statistical methods are developed and implemented in the procedures. Each new version of the software contains new capabilities. The survey analysis procedures currently perform many of the most common types of survey data analyses. The use of replicate weights allows analyses to be done with other SAS software procedures on survey data, as we saw in Section 10.3 for fitting loglinear models. A similar macro can be used to perform other analyses not yet implemented in the survey software. For example, standard errors for quantile regression coefficients and predictions can be calculated through a macro by using the balanced repeated replication or bootstrap method.

Numerous user-contributed macros add to the capabilities of the software. You can find recent contributions in the annual SAS Global Forum proceedings, available for free download from `https://support.sas.com`. SAS® Users Groups, which are informal associations of persons who are interested in SAS software (see `https://www.sas.com/en_us/connect/user-groups.html`), provide another resource for learning about new procedures and macros.

Still have questions after reading this book? Check out the documentation, technical tips, training, and communities at `support.sas.com`. The chances are that you are not the first to have your question, and this website likely contains an answer to it.

A

Data Set Descriptions

The data sets referenced in SDA and described in this appendix are available from the book website (see the Preface for the website address) and in the contributed R package SDAResources (Lu and Lohr, 2021). These data sets are provided for instructional purposes only and without warranty. Anyone wishing to investigate the subject matter further should obtain the original data from the source. In some cases, the data sets referenced in SDA and this book are a subset of the original data; in others, the information has been modified to protect the confidentiality of the respondents.

All data sets ending in `.csv` use commas as a separator between fields.

These data sets have also been stored in SAS format (with the name ending in .sas7bdat) and R format with missing values recoded to the symbols used for missing data in the software package ('.' or blank in SAS and NA in R).

agpop.csv Data from the 1992 U.S. Census of Agriculture. Source: U.S. Bureau of the Census (1995). In columns 3–14, the value −99 denotes missing data.

Column	Name	Value
1	county	county name (character variable)
2	state	state abbreviation (character variable)
3	acres92	number of acres devoted to farms, 1992
4	acres87	number of acres devoted to farms, 1987
5	acres82	number of acres devoted to farms, 1982
6	farms92	number of farms, 1992
7	farms87	number of farms, 1987
8	farms82	number of farms, 1982
9	largef92	number of farms with 1,000 acres or more, 1992
10	largef87	number of farms with 1,000 acres or more, 1987
11	largef82	number of farms with 1,000 acres or more, 1982
12	smallf92	number of farms with 9 acres or fewer, 1992
13	smallf87	number of farms with 9 acres or fewer, 1987
14	smallf82	number of farms with 9 acres or fewer, 1982
15	region	S = south; W = west; NC = north central; NE = northeast

agpps.csv Data from a without-replacement probability-proportional-to-size sample from file `agpop.csv`.

DOI: 10.1201/9781003160366-A

Column	Name	Value
1	county	county name
2	state	state abbreviation
3	acres92	number of acres devoted to farms, 1992
4	acres87	number of acres devoted to farms, 1987
5–15	...	same as variables 5–15 in `agpop.csv`
16	sizemeas	size measure used to select the pps sample
17	SelectionProb	inclusion probability for county i, π_i
18	SamplingWeight	sampling weight for county i, $w_i = 1/\pi_i$
19	Unit	unit number for indexing joint inclusion probabilities
20–34	JtProb_1–JtProb_15	columns of joint inclusion probabilities

agsrs.csv Data from an SRS of size 300 from the 1992 U.S. Census of Agriculture. Variables are the same as in `agpop.csv`. In columns 3–14, the value −99 denotes missing data.

agstrat.csv Data from a stratified random sample of size 300 from the 1992 U.S. Census of Agriculture data in `agpop.csv`. In columns 3–14, the value −99 denotes missing data.

Column	Name	Value
1	county	county name
2	state	state abbreviation
3	acres92	number of acres devoted to farms, 1992
4	acres87	number of acres devoted to farms, 1987
5	acres82	number of acres devoted to farms, 1982
6	farms92	number of farms, 1992
7	farms87	number of farms, 1987
8	farms82	number of farms, 1982
9	largef92	number of farms with 1,000 acres or more, 1992
10	largef87	number of farms with 1,000 acres or more, 1987
11	largef82	number of farms with 1,000 acres or more, 1982
12	smallf92	number of farms with 9 acres or fewer, 1992
13	smallf87	number of farms with 9 acres or fewer, 1987
14	smallf82	number of farms with 9 acres or fewer, 1982
15	region	S = south; W = west; NC = north central; NE = northeast
16	rn	random numbers used to select sample in each stratum
17	strwt	sampling weight for each county in sample

algebra.csv Hypothetical data for an SRS of 12 algebra classes in a city, from a population of 187 classes.

Column	Name	Value
1	class	Class number
2	Mi	Number of students (M_i) in class
3	score	Score of student on test

anthrop.csv Finger length and height for 3,000 criminals. Source: Macdonell (1901). This data set contains information for the entire population.

Column	Name	Value
1	finger	length of left middle finger (cm)
2	height	height (inches)

anthsrs.csv Finger length and height for an SRS of size 200 from anthrop.csv.

Column	Name	Value
1	finger	length of left middle finger (cm)
2	height	height (inches)
3	wt	sampling weight

anthuneq.csv Finger length and height for a with-replacement unequal-probability sample of size 200 from `anthrop.csv`. The probability of selection, ψ_i, was proportional to 24 for $y < 65$, 12 for $y = 65$, 2 for $y = 66$ or 67, and 1 for $y > 67$.

Column	Name	Value
1	finger	length of left middle finger (cm)
2	height	height (inches)
3	wt	sampling weight

artifratio.csv Values from all possible SRSs for artificial population in Chapter 4 of SDA.

Column	Name	Value
1	sample	sample number
2	i1	first unit in sample
3	i2	second unit in sample
4	i3	third unit in sample
5	i4	fourth unit in sample
6	xbars	$\bar{x}_{\mathcal{S}}$
7	ybars	$\bar{y}_{\mathcal{S}}$
8	bhat	$\hat{B}$
9	tSRS	$\hat{t}_{y,\text{SRS}} = N\bar{y}_{\mathcal{S}}$
10	thatr	$\hat{t}_{yr}$

asafellow.csv Information from a stratified random sample of Fellows of the American Statistical Association elected between 2000 and 2018. The list of Fellows serving as the population was downloaded from `https://www.amstat.org/ASA/Your-Career/Awards/ASA-Fellows-list.aspx` on March 18, 2019. All other information was obtained from public sources.

Column	Name	Value
1	awardyr	Year of award
2	gender	Gender of Fellow (character variable, M = male, F = female)

asafellow.csv (continued)

Column	Name	Value
3	popsize	Population size in stratum (= N_h)
4	sampsize	Sample size in stratum (= n_h)
5	field	Field of employment (character variable) acad = academia, ind = industry, govt = government
6	degreeyr	Year in which Fellow received terminal degree (year of Ph.D. if applicable, otherwise year of Master's or Bachelor's degree)
7	math	= 1 if majored in mathematics as undergraduate, 0 if did not major in math, −99 if missing

auditresult.csv Audit data used in Chapter 6 of SDA.

Column	Name	Value
1	account	audit unit
2	bookvalue	book value of account
3	psi	probability of selection
4	auditvalue	audit value of account

auditselect.csv Selection of accounts for audit data used in Chapter 6 of SDA.

Column	Name	Value
1	account	audit unit
2	bookval	book value of account
3	cumbv	cumulative book value
4	rn1	random number 1 selecting account
5	rn2	random number 2 selecting account
6	rn3	random number 3 selecting account

azcounties.csv Population and housing unit estimates for Arizona counties (excluding Maricopa and Pima counties), from the American Community Survey 2018 5-year estimates. Source: `https://data.census.gov/`, accessed November 27, 2020.

Column	Name	Value
1	number	County number
2	name	County name (character variable, length 15)
3	population	Population estimate for county
4	housing	Housing unit estimate for county
5	ownerocc	Number of owner-occupied housing units for county

baseball.csv Statistics on 797 baseball players, compiled by Jenifer Boshes from the rosters of all major league teams in November 2004. Source: Forman (2004). Missing values (for variables *pball, intwalk, hbp,* and *sacrfly*; all other variables have complete data) are coded as −9.

Column	Name	Value
1	team	team played for at beginning of the season
2	leagueid	AL or NL
3	player	a unique identifier for each baseball player
4	salary	player salary in 2004
5	pos	primary position coded as P, C, 1B, 2B, 3B, SS, RF, LF, or CF
6	gplay	games played
7	gstart	games started
8	inning	number of innings
9	putout	number of putouts
10	assist	number of assists
11	error	Errors
12	dplay	number of double plays
13	pball	number of passed balls (only applies to catchers)
14	gbat	number of games that player appeared at bat
15	atbat	number of at bats
16	run	number of runs scored
17	hit	number of hits
18	secbase	number of doubles
19	thirdbase	number of triples
20	homerun	number of home runs
21	rbi	number of runs batted in
22	stolenb	number of stolen bases
23	csteal	number of times caught stealing
24	walk	number of times walked
25	strikeout	number of strikeouts
26	intwalk	number of times intentionally walked
27	hbp	number of times hit by pitch
28	sacrhit	number of sacrifice hits
29	sacrfly	number of sacrifice flies
30	gidplay	grounded into double play

books.csv Data from homeowner's survey to estimate total number of books, used in Chapter 5.

Column	Name	Value
1	shelf	shelf number
2	Mi	number of books on shelf
3	booknumber	number of the book selected
4	purchase	purchase cost of book
5	replace	replacement cost of book

census1920.csv Population sizes for each state, from the 1920 U.S. census. The data set contains only the 48 states and excludes Washington D.C., Puerto Rico, and U.S. territories (these areas were not allowed to have voting representatives in Congress). Source: U.S. Bureau of the Census (1921).

Column	Name	Value
1	state	state name
2	population	state population in 1920 census

census2010.csv Population sizes for each state, from the 2010 U.S. census. Source: U.S. Census Bureau (2019). The data set contains only the 50 states and excludes the areas that, as of 2010, were not allowed to have voting representatives in Congress: Washington D.C., Puerto Rico, and U.S. territories.

Column	Name	Value
1	state	state name
2	population	state population in 2010 census

cherry.csv Data for a sample of 31 cherry trees. Source: Hand et al. (1994).

Column	Name	Value
1	diameter	Diameter of tree (inches)
2	height	Height of tree (feet)
3	volume	Timber volume of tree (cubic feet)

classes.csv Population sizes for 15 classes, used in Chapter 6 of SDA to illustrate unequal-probability sampling.

Column	Name	Value
1	class	Class ID number
2	class_size	Number of students in class

classpps.csv Two-stage unequal-probability sample without replacement from the population of classes in `classes.csv`.

Column	Name	Value
1	class	Class ID number
2	class_size	Number of students in class
3	finalweight	Sampling weight for student
4	hours	Number of hours spent studying statistics

classppsjp.csv Joint inclusion probabilities for unequal-probability sample without replacement from the population of classes in `classes.csv`.

Column	Name	Value
1	class	Class ID number
2	class_size	Number of students in class
3	SelectionProb	Probability of being included in sample, π_i
4	SamplingWeight	Sampling weight $w_i = 1/\pi_i$
5–9	JtProb_1–JtProb_5	Columns of joint inclusion probabilities, π_{ik}

college.csv Selected variables from the U.S. Department of Education College Scorecard Data (version updated on June 1, 2020). Source: U.S. Department of Education (2020), downloaded on August 25, 2020. Some of the variables in `college.csv` have been calculated from other variables in the original source; these have been given new variable names that are not found in the data dictionary at `https://collegescorecard.ed.gov/data/documentation/`.

This data set is made available for pedagogical purposes only. Anyone wishing to draw conclusions from College Scorecard data should obtain the full data set from the Department of Education. The original data set has 1,925 variables and includes institutions (such as those that do not grant undergraduate degrees) that are not in the file `college.csv`.

The data set `college.csv` includes institutions in the original data set that: (1) are located in the 50 states plus the District of Columbia, (2) contain information on average net price (NPT4), (3) are predominantly Bachelor's degree-granting, (4) were currently operating as of June 2020, (5) are not private for-profit institutions or "global" campuses, (6) have Carnegie size classification (variable *ccsizset*) between 6 and 17 and Carnegie basic classification (variable *ccbasic*) between 14 and 22 (these offer Bachelor's degrees), (7) enroll first-time students, and (8) are not U.S. Service Academies.

For all variables, missing data are coded as −99.

Column	Name	Value
1	unitid	Unit identification number
2	instnm	Institution name (character, length 81)
3	city	City (character, length 24)
4	stabbr	State abbreviation (character, length 2)
5	highdeg	Highest degree awarded 3 = Bachelor's degree, 4 = Graduate degree
6	control	Control (ownership) of institution 1 = Public, 2 = Private nonprofit
7	region	Region where institution is located 1 New England (CT, ME, MA, NH, RI, VT) 2 Mid East (DE, DC, MD, NJ, NY, PA) 3 Great Lakes (IL, IN, MI, OH, WI) 4 Plains (IA, KS, MN, MO, NE, ND, SD) 5 Southeast (AL, AR, FL, GA, KY, LA, MS, NC, SC, TN, VA, WV) 6 Southwest (AZ, NM, OK, TX) 7 Rocky Mountains (CO, ID, MT, UT, WY) 8 Far West (AK, CA, HI, NV, OR, WA)
8	locale	Locale of institution 11 City: Large (population of 250,000 or more) 12 City: Midsize (population of at least 100,000 but less than 250,000) 13 City: Small (population less than 100,000) 21 Suburb: Large (outside principal city, in urbanized area with population of 250,000 or more) 22 Suburb: Midsize (outside principal city, in urbanized area with population of at least 100,000 but less than 250,000)

college.csv (continued)

Column	Name	Value
		23 Suburb: Small (outside principal city, in urbanized area with population less than 100,000) 31 Town: Fringe (in urban cluster up to 10 miles from an urbanized area) 32 Town: Distant (in urban cluster more than 10 miles and up to 35 miles from an urbanized area) 33 Town: Remote (in urban cluster more than 35 miles from an urbanized area) 41 Rural: Fringe (rural territory up to 5 miles from an urbanized area or up to 2.5 miles from an urban cluster) 42 Rural: Distant (rural territory more than 5 miles but up to 25 miles from an urbanized area or more than 2.5 and up to 10 miles from an urban cluster) 43 Rural: Remote (rural territory more than 25 miles from an urbanized area and more than 10 miles from an urban cluster)
9	ccbasic	Carnegie basic classification 15 Doctoral Universities: Very High Research Activity 16 Doctoral Universities: High Research Activity 17 Doctoral/Professional Universities 18 Master's Colleges & Universities: Larger Programs 19 Master's Colleges & Universities: Medium Programs 20 Master's Colleges & Universities: Small Programs 21 Baccalaureate Colleges: Arts & Sciences Focus 22 Baccalaureate Colleges: Diverse Fields
10	ccsizset	Carnegie classification, size and setting 6 Four-year, very small, primarily nonresidential 7 Four-year, very small, primarily residential 8 Four-year, very small, highly residential 9 Four-year, small, primarily nonresidential 10 Four-year, small, primarily residential 11 Four-year, small, highly residential 12 Four-year, medium, primarily nonresidential 13 Four-year, medium, primarily residential 14 Four-year, medium, highly residential 15 Four-year, large, primarily nonresidential 16 Four-year, large, primarily residential 17 Four-year, large, highly residential
11	hbcu	Historically black college or university, 1 = yes, 0 = no
12	openadmp	Does the college have an open admissions policy, that is, does it accept any students that apply or have minimal requirements for admission? 1 = yes, 0 = no
13	adm_rate	Fall admissions rate, defined as the number of admitted undergraduates divided by the number of undergraduates who applied
14	sat_avg	Average SAT score (or equivalent) for admitted students
15	ugds	Number of number of degree-seeking undergraduate students enrolled in the fall term

college.csv (continued)

Column	Name	Value
16	ugds_men	Proportion of *ugds* who are men
17	ugds_women	Proportion of *ugds* who are women
18	ugds_white	Proportion of *ugds* who are white (based on self-reports)
19	ugds_black	Proportion of *ugds* who are black/African American (based on self-reports)
20	ugds_hisp	Proportion of *ugds* who are Hispanic (based on self-reports)
21	ugds_asian	Proportion of *ugds* who are Asian (based on self-reports)
22	ugds_other	Proportion of *ugds* who have other race/ethnicity (created from other categories on original data file; race/ethnicity proportions sum to 1)
23	npt4	Average net price of attendance, derived from the full cost of attendance (including tuition and fees, books and supplies, and living expenses) minus federal, state, and institutional grant/scholarship aid, for full-time, first-time undergraduate Title IV-receiving students. NPT4 created from scorecard data variables NPT4_PUB if public institution and NPT4_PRIV if private
24	tuitionfee_in	In-state tuition and fees
25	tuitionfee_out	Out-of-state tuition and fees
26	avgfacsal	Average faculty salary per month
27	pftfac	Proportion of faculty that is full-time
28	c150_4	Proportion of first-year, full-time students who complete their degree within 150% of the expected time to complete; for most institutions, this is the proportion of students who receive a degree within 6 years
29	grads	Number of graduate students

collegerg.csv Five replicate SRSs from the set of public colleges and universities (having *control* = 1) in `college.csv`. Columns 1–29 are as in `college.csv`, with additional columns 30–32 listed below. Note that the selection probabilities and sampling weights are for the separate replicate samples, so that the weights for each replicate sample sum to the population size 500.

Column	Name	Value
30	selectionprob	Selection probability for each replicate sample
31	samplingweight	Sampling weight for each replicate sample
32	repgroup	Replicate group number

collshr.csv Probability-proportional-to-size sample of size 10 from the stratum of small, highly residential colleges (having *ccsizeset* = 11) in `college.csv`. Columns 1–29 are as in `college.csv`, with additional columns 30–34 listed below.

Column	Name	Value
30	mathfac	Number of mathematics faculty
31	psychfac	Number of psychology faculty
32	biolfac	Number of biology faculty
33	psii	Selection probability, $= ugds/(\text{sum of } ugds \text{ for stratum})$
34	wt	Sampling weight $= 1/(10\psi_i)$

coots.csv Selected information on egg size, from a larger study by Arnold (1991). Data provided courtesy of Todd Arnold. Not all observations are used for this data set, so results may not agree with those in Arnold (1991).

Column	Name	Value
1	clutch	clutch number from which eggs were subsampled.
2	csize	number of eggs in clutch (M_i)
3	length	length of egg (mm)
4	breadth	maximum breadth of egg (mm)
5	volume	calculated as 0.000507*length * breadth^2 (mm^3)
6	tmt	= 1 if received supplemental feeding, 0 otherwise

counties.csv Data (from 1990) from an SRS of 100 of the 3141 counties in the United States. Missing values are coded as −99. Source: U.S. Census Bureau (1994).

Column	Name	Value
1	RN	random number used to select the county
2	state	state abbreviation
3	county	county name
4	landarea	land area, 1990 (square miles)
5	totpop	total number of persons, 1992
6	physician	active non-Federal physicians on Jan. 1, 1990
7	enroll	school enrollment in elementary or high school, 1990
8	percpub	percent of school enrollment in public schools
9	civlabor	civilian labor force, 1991
10	unemp	number unemployed, 1991
11	farmpop	farm population, 1990
12	numfarm	number of farms, 1987
13	farmacre	acreage in farms, 1987
14	fedgrant	total expenditures in federal funds and grants, 1992 (millions of dollars)
15	fedciv	civilians employed by federal government, 1990
16	milit	military personnel, 1990
17	veterans	number of veterans, 1990
18	percviet	percent of veterans from Vietnam era, 1990

crimes.csv Data from selected variables in a simple random sample of 5,000 records from the 7,048,107 records with dates between 2001 and 2019 in the City of Chicago database "Crimes—2001 to Present." This file was downloaded on August 11, 2020 from `https://data.cityofchicago.org/`. These data are provided for pedagogical purposes only. Anyone wishing to publish analyses of Chicago crime data should obtain the most recent data from `https://data.cityofchicago.org/`. For a list and map of Community Areas, see `https://www.chicago.gov/city/en/depts/dgs/supp_info/citywide_maps.html`.

Column	Name	Value
1	year	Year in which crime occurred (between 2001 and 2019)
2	crimetype	Type of crime, determined from detailed crime description in database homicide = homicide, sexualasslt = sexual assault, robbery = robbery, aggasslt = aggravated assault, burglary = burglary, mvtheft = motor vehicle theft, idtheft = identity theft, theft = other type of theft, arson = arson, simpleasslt = simple assault (assaults that are not aggravated), threat = threat or harassment, fraud = fraud, weapon = weapons violation, trespass = trespassing, narcotics = narcotics or liquor law violation, vandalism = vandalism, other = other
3	violent	= 1 if violent crime, 0 otherwise
4	arrest	= 1 if an arrest was made, 0 otherwise
5	domestic	= 1 if crime was domestic-related as defined by the Illinois Domestic Violence Act, 0 otherwise
6	commarea	Number of the Community Area in Chicago where the crime occurred
7	location	Type of location where crime occurred (e.g., street, apartment)

deadtrees.csv Number of dead trees recorded by photograph and field count for a (fictional) SRS of 25 plots taken from a population of 100 plots.

Column	Name	Value
1	photo	Number of dead trees in plot from photograph
2	field	Number of dead trees in plot from field observation

divorce.csv Data from a sample of divorce records for states in the Divorce Registration Area. Source: National Center for Health Statistics (1987).

Column	Name	Value
1	state	state name (character variable)
2	abbrev	state abbreviation (character variable)
3	samprate	sampling rate for state
4	numrecs	number of records sampled in state
5	hsblt20	number of records in sample with husband's age < 20
6	hsb20to24	number of records with $20 \leq$ husband's age ≤ 24
7	hsb25to29	number of records with $25 \leq$ husband's age ≤ 29
8	hsb30to34	number of records with $30 \leq$ husband's age ≤ 34
9	hsb35to39	number of records with $35 \leq$ husband's age ≤ 39
10	hsb40to44	number of records with $40 \leq$ husband's age ≤ 44

divorce.csv (continued)

Column	Name	Value
11	hsb45to49	number of records with 45 $\leq$ husband's age $\leq$ 49
12	hsbge50	number of records with husband's age $\geq$ 50
13	wflt20	number of records with wife's age $<$ 20
14	wf20to24	number of records with 20 $\leq$ wife's age $\leq$ 24
15	wf25to29	number of records with 25 $\leq$ wife's age $\leq$ 29
16	wf30to34	number of records with 30 $\leq$ wife's age $\leq$ 34
17	wf35to39	number of records with 35 $\leq$ wife's age $\leq$ 39
18	wf40to44	number of records with 40 $\leq$ wife's age $\leq$ 44
19	wf45to49	number of records with 45 $\leq$ wife's age $\leq$ 49
20	wfge50	number of records with wife's age $\geq$ 50

gini.csv Data from the population of districts for the 1921 Italian general census. Source: Gini and Galvani (1929, pp. 73–78).

Column	Name	Value
1	id	ID number
2	district	District name
3	birth_rate	Births per 1,000 population
4	death_rate	Deaths per 1,000 population
5	marriage_rate	Marriages per 1,000 population
6	agricultural_pop	Percentage of males over 10 years old who work in agriculture
7	urban_population	Percentage of population in urban areas
8	income	Average income
9	altitude	Average altitude above sea level (meters)
10	pop_density	Number of inhabitants per square kilometer
11	natural_growth	Rate of average increase of the population
12	population	Population of area
13	area	Land area (square kilometers)
14	in_GG_sample	= 1 if in the purposive sample selected by Gini and Galvani; 0 otherwise

golfsrs.csv A simple random sample of 120 golf courses, taken from the population on the website `ww2.golfcourse.com` on August 5, 1998. Missing data in the `.csv` file are denoted by blanks.

Column	Name	Value
1	RN	random number used to select golf course for sample
2	state	state name
3	holes	number of holes
4	type	type of course: priv = private, semi = semi-private, pub = public, mili = military, resort
5	yearblt	year course was built
6	wkday18	greens fee for 18 holes during week
7	wkday9	greens fee for 9 holes during week
8	wkend18	greens fee for 18 holes on weekend

golfsrs.csv (continued)

Column	Name	Value
9	wkend9	greens fee for 9 holes on weekend
10	backtee	back tee yardage
11	rating	course rating
12	par	par for course
13	cart18	golf cart rental fee for 18 holes
14	cart9	golf cart rental fee for 9 holes
15	caddy	Are caddies available? (y or n)
16	pro	Is a golf pro available? (y or n)

gpa.csv GPA data from Chapter 5 of SDA.

Column	Name	Value
1	suite	Suite (psu) identifier
2	gpa	Grade point average of person in suite
3	wt	Sampling weight, = 20 for every observation

healthjournals.csv Randomization and statistical inference practices in a stratified random sample of 196 public health articles. The data, provided courtesy of Dr. Matt Hayat, are discussed in Hayat and Knapp (2017). The variables provided in `healthjournals.csv` are a subset of the variables collected by the authors.

Column	Name	Value
1	journal	Journal that published the article AJPH = *American Journal of Public Health* AJPM = *American Journal of Preventive Medicine* PM = *Preventive Medicine*
2	NumAuthors	Number of authors
3	RandomSel	= "Yes" if data in the article were from a randomly selected (probability) sample; "No" otherwise
4	RandomAssn	= "Yes" if study subjects for the article were randomly assigned to treatment groups; "No" otherwise
5	ConfInt	= "Yes" if a confidence interval appeared in the article's main text, tables, or figures; "No" otherwise
6	HypTest	= "Yes" if a p-value or significance test appeared in the article's main text, tables, or figures; "No" otherwise
7	Asterisks	= "Yes" if asterisks were used to represent p-value ranges; "No" otherwise

htcdf.csv Empirical distribution function and empirical probability mass function of data in `htpop.csv`.

Column	Name	Value
1	height	height value, cm
2	frequency	number of times height value in column 1 occurs in population
3	epmf	empirical probability mass function
4	ecdf	empirical distribution function

htpop.csv Height and gender of 2,000 persons in an artificial population.

Column	Name	Value
1	height	height of person, cm
2	gender	M=male, F=female

htsrs.csv Height and gender for a SRS of 200 persons, taken from `htpop.csv`.

Column	Name	Value
1	rn	random number used to select unit
2	height	height of person, cm
3	gender	M=male, F=female

htstrat.csv Height and gender for a stratified random sample of 160 women and 40 men, taken from `htpop.csv`. The columns and names are as in `htsrs.csv`.

hunting.csv Population and sample sizes for the poststrata used for the Sunday hunting survey. Source: Virginia Polytechnic and State University/Responsive Management (2006).

Column	Name	Value
1	region	Region of state (East, Central, West)
2	gender	Gender (female, male)
3	age	Age group (16-24, 25-34, 35-44, 45-54, 55-64, 65+)
4	popsize	Population size in poststratum from the 2000 U.S. census
5	sampsize	Sample size in poststratum

impute.csv Small artificial data set used to illustrate imputation methods. Missing values are denoted by −99.

Column	Name	Value
1	person	identification number for person
2	age	age in years
3	gender	M=male, F=female
4	education	number of years of education
5	crime	= 1 if victim of any crime, 0 otherwise
6	violcrime	= 1 if victim of violent crime, 0 otherwise

integerwt.csv Artificial population of 2,000 observations.

Column	Name	Value
1	stratum	Stratum number
2	y	y value of observation

intellonline.csv Data from the online (Mechanical Turk) survey. Source: Heck et al. (2018). The data were downloaded from `https://journals.plos.org/plosone/article?id=10.1371/journal.pone.0200103` on February 8, 2020; the variables extracted from the full data set are provided here for educational purposes only.

Column	Name	Value
1	int	Response to question about agreement with the statement "I am more intelligent than the average person." 1 = Strongly Agree; 2 = Mostly Agree; 3 = Mostly Disagree; 4 = Strongly Disagree; 5 = Don't Know or Not Sure
2	region	Census region of respondent (character variable, length 10): Northeast, South, Midwest, West
3	sex	Sex (character variable, length 8): Male, Female
4	race	Race (character variable, length 18): White, African American, Asian American, Hispanic American, Another origin
5	age	Age, years
6	income	Household income level (character variable, length 8): <$40k, $40–80k, or >$80k
7	education	Highest education level attained (character variable, length 12): No College, Some College, College Grad, Grad School
8	postwt	Relative weight, obtained by poststratifying to demographic proportions in the 2010 U.S. Census. The weights are normed so that they sum to 750.

intelltel.csv Data from the telephone survey studied by Heck et al. (2018). The data were downloaded from `https://journals.plos.org/plosone/article?id=10.1371/journal.pone.0200103` and are provided here for educational purposes only. The variables are the same as in `intellonline.csv`.

intellwts.csv Relative weights for demographic groups in `intellonline.csv` and `intelltel.csv` (Heck et al., 2018). Each sample was weighted using the 2010 U.S. Census demographics for sex (male, female), age (< 44, ≥ 44), and race/ethnicity (white, nonwhite). The table entries give the weights for each of these eight demographic groups.

Column	Name	Value
1	sex	Sex
2	agegroup	Age group: Young = (age less than 44), Old = (age greater than or equal to 44)
3	race	Race: White or Nonwhite
4	tel_n	Number of telephone survey respondents in the sex/age-group/race class

intellwts.csv (continued)

Column	Name	Value
5	online_n	Number of online survey respondents in the sex/agegroup/race class
6	tel_wgt	Relative weight for each respondent to the telephone survey in this sex/agegroup/race class
7	online_wgt	Relative weight for each respondent to the telephone survey in this sex/agegroup/race class

ipums.csv Data extracted from the 1980 Census Integrated Public Use Microdata Series, using the "Small Sample Density" option in the data extract tool, on September 17, 2008. The stratum and psu variables were constructed for use in the book exercises. Data analyses on this file do NOT give valid results for inference to the 1980 U.S. population. Source: Ruggles et al. (2004).

Column	Name	Value
1	stratum	stratum number (1–9)
2	psu	psu number (1–90)
3	inctot	total personal income (dollars), topcoded at $75,000
4	age	age, with range 15–90
5	sex	1 = Male, 2 = Female
6	race	1 = White, 2 = Black, 3 = American Indian or Alaska Native, 4 = Asian or Pacific Islander, 5 = Other Race
7	hispanic	0 = Not Hispanic, 1 = Hispanic
8	marstat	Marital Status: 1 = Married, 2 = Separated, 3 = Divorced, 4 = Widowed, 5 = Never married/single
9	ownershg	Ownership of housing unit: 0 = Not Applicable (N/A), 1 = Owned or being bought, 2 = Rents
10	yrsusa	Number of years a foreign-born person has lived in the U.S.: 0= N/A, 1= 0–5 years, 2= 6–10 years, 3= 11–15 years, 4= 16–20 years, 5= 21+ years
11	school	Is person in school? 0 = N/A, 1 = No, not in school, 2 = Yes, in school
12	educrec	Educational Attainment: 1= None or preschool, 2= Grade 1, 2, 3, or 4, 3= Grade 5, 6, 7, or 8, 4= Grade 9, 5= Grade 10, 6= Grade 11, 7= Grade 12, 8= 1 to 3 years of college, 9= 4+ years of college
13	labforce	In labor force? 0 = Not Applicable, 1 = No, 2 = Yes
14	classwk	class of worker: 0=Not applicable, 13= Self-employed, not incorporated, 14= Self-employed, incorporated, 22= Wage/salary, private, 25= Federal government employee, 27= State government employee, 28= Local government employee, 29= Unpaid family worker
15	vetstat	Veteran Status 0 = Not Applicable, 1 = No Service, 2 = Yes

journal.csv Types of sampling used for articles in a sample of journals. Source: Jacoby and Handlin (1991).

Note that columns 2 and 3 do not always sum to column 1; for some articles, the investigators could not determine which type of sampling was used. When working with these data, you may wish to create a fourth column, "indeterminate," which equals column1 − (column2 + column3).

Column	Name	Value
1	numemp	number of articles in 1988 that used sampling
2	prob	number of articles that used probability sampling
3	nonprob	number of articles that used non-probability sampling

measles.csv Roberts et al. (1995) reported on the results of a survey of parents whose children had not been immunized against measles during a recent campaign to immunize all children in the first five years of secondary school. The original data were unavailable; univariate and multivariate summary statistics from these artificial data, however, are consistent with those in the paper. All variables are coded as 1 for yes, 0 for no, and 9 for no answer. A parent who refused consent (variable 4) was asked why, with responses in variables 5 through 10. If a response in variables 5 through 10 was checked, it was assigned value 1; otherwise, it was assigned value 0. A parent could give more than one reason for not having the child immunized.

Column	Name	Value
1	school	school attended by child
2	form	Parent received consent form
3	returnf	Parent returned consent form
4	consent	Parent gave consent for measles immunization
5	hadmeas	Child had already had measles
6	previmm	Child had been immunized against measles
7	sideeff	Parent concerned about side effects
8	gp	Parent wanted general practitioner (GP) to give vaccine
9	noshot	Child did not want injection
10	notser	Parent thought measles was not a serious illness
11	gpadv	GP advised that vaccine was not needed
12	Mitotal	Population size in school
13	mi	Sample size in school

mysteries.csv Data from a stratified random sample of books nominated for the Edgar® awards for Best Novel and Best First Novel. The sample was drawn from the population listing of 655 books at `http://theedgars.com/awards/` on August 14, 2020.

Column	Name	Value
1	stratum	Stratum number, from 1 to 12, computed from the stratification variables in columns 2–4
2	time	Time period in which award was given: 1 = 1946–1980, 2 = 1981–2000, 3 = 2001–2020
3	category	Award category (character variable, length 16): Best Novel, or Best First Novel

mysteries.csv (continued)

Column	Name	Value
4	winner	= 1 if book won the award that year, = 0 if book was nominated but did not win award
5	popsize	Number of population books in stratum ($= N_h$)
6	sampsize	Number of sampled books in stratum ($= n_h$)
7	obtained	= 1 if book was obtained (responded) in original sample, = 2 if book was obtained in phase II subsample of nonrespondents, = 0 if not obtained
8	p1weight	Weight for phase I sample, calculated as N_h/n_h; use for exercises in Chapters 1–11 of SDA
9	p2weight	Final weight for phase II sample; use for exercises in Chapter 12 of SDA and analyses involving variables *victims* and *firearm*
10	genre	Genre of book (character variable, length 11). Values "private eye" (protagonist is a private detective), "procedural" (a detailed, step-by-step analysis of how the crime is solved, using the skills of the detective), or "suspense" (the protagonist is at the center of action or is involved in espionage, but is not a professional detective)
11	historical	= 1 if the main action in the book takes place at least 20 years before the book's publication date, = 0 if book action is within 20 years of the publication date
12	urban	= 1 if the main action in the book takes place primarily in urban areas, = 0 otherwise
13	authorgender	Gender of author (character variable, length 1) = "F" if author is female, "M" if author is male
14	fdetect	Number of female detectives (or protagonists, if book has no detective) in book
15	mdetect	Number of male detectives (or protagonists, if book has no detective) in book
16	victims	Number of murder victims in book (missing value set to −9 if *obtained* = 0)
17	firearm	Number of murders committed with firearms in book (missing value set to −9 if *obtained* = 0)

nhanes.csv Selected variables from the 2015–2016 National Health and Nutrition Examination Survey (NHANES). Source: Centers for Disease Control and Prevention (2017). This data set is provided for educational purposes only. Anyone wishing to publish or use results from analyses of NHANES data should obtain the data files directly from the source.

The data files merged to create `nhanes.csv` can be read directly from the SAS transport files `DEMO_I.XPT`, `BMX_I.XPT`, `TCHOL_I.XPT`, and `BPX_I.XPT` from the NHANES website. Variables 1–23 have the same names as in the SAS transport files.

The blood pressure variables *sbp* and *dbp* were created as follows. In the medical examination, three consecutive blood pressure readings were obtained after participants sat quietly for 5 minutes, and the maximum inflation level was determined. A fourth measurement was conducted for some persons who had an incomplete or interrupted blood pressure reading.

The variables *sbp* and *dbp* were calculated by discarding the first blood pressure reading and calculating the average of the remaining valid readings. Note that some of the diastolic blood pressure readings are 0.

In the comma-delimited file `nhanes.csv`, missing values are denoted by −9. In the SAS data file, missing values are denoted by a period. In the R data file, missing values are denoted by NA. Note that some of the codes for variables in the table below also denote missing values; for example, the value 7 for *dmdeduc2* indicates "Refused," and these codes for special types of missing values remain in the SAS and R data files.

Column	Name	Value
1	sdmvstra	Pseudo-stratum. These are groups of secondary sampling units used for variance estimation on the publicly available data. Pseudo-strata and pseudo-psus are released instead of the actual strata and psus to protect the confidentiality of respondents' information. Use *sdmvstra* as the variable defining the strata.
2	sdmvpsu	Pseudo-psu. Use *sdmvpsu* as the primary sampling unit (psu). There are two pseudo-psus per pseudo-stratum, numbered 1 and 2.
3	wtint2yr	Interview weight (use as weight for variables 5–12)
4	wtmec2yr	Mobile Examination Center weight (use as weight for any analysis involving variables 13–25)
5	ridstatr	Interview/examination status, = 1 if interviewed only, = 2 if interviewed and had medical examination
6	ridageyr	Age in years at screening, from 0 to 80. Anyone with age > 80 years is recorded (topcoded) as 80. No values are missing for this variable.
7	ridagemn	Age in months at screening (reported only for persons with age 24 months or younger at the time of exam, otherwise missing)
8	riagendr	= 1 if male, 2 if female (no missing values)
9	ridreth3	Race/ethnicity code (no missing values) 1 = Mexican American 2 = Other Hispanic 3 = Non-Hispanic White 4 = Non-Hispanic Black 6 = Non-Hispanic Asian 7 = Other Race, Including Multi-Racial
10	dmdeduc2	Education level of person interviewed (given for adults age 20+ only) 1 = Less than 9th grade 2 = 9th to 11th grade (including 12th grade with no diploma) 3 = High school graduate (including GED) 4 = Some college or associate's degree 5 = College graduate or above 7 = Refused 9 = Don't know
11	dmdfmsiz	Total number of people in the family. Values 1–6 indicate the number of people is that number; value 7 indicates 7 or more people in family. No missing values.

nhanes.csv (continued)

Column	Name	Value
12	indfmpir	Ratio of family income to poverty guideline. A value less than 1 indicates the family is below the poverty threshold. Variable *indfmpir* is a continuous variable where values between 0 and 4.99 indicate the actual poverty ratio. A value of 5 indicates that the ratio of family income to the poverty guideline for that family is 5 or more.
13	bmxwt	Weight (kg)
14	bmxht	Standing height (cm)
15	bmxbmi	Body mass index (kg/m^2), calculated as bmxwt/(bmxht/100)^2
16	bmxwaist	Waist circumference (cm)
17	bmxleg	Upper leg length (cm)
18	bmxarml	Upper arm length (cm)
19	bmxarmc	Upper arm circumference (cm)
20	bmdavsad	Average sagittal abdominal diameter (SAD, the distance from the small of the back to the upper abdomen), in cm. Calculated by averaging the SAD readings on the person (up to four).
21	lbxtc	Serum total cholesterol (mg/dL)
22	bpxpls	60-second pulse
23	sbp	Average systolic blood pressure (mm Hg)
24	dbp	Average diastolic blood pressure (mm Hg)
25	bpread	Number of blood pressure readings

nybight.csv Data collected in the New York Bight for June 1974 and June 1975. Two of the original strata were combined because of insufficient sample sizes. For variable *catchwt*, weights less than 0.5 were recorded as 0.5 kg. Source: Wilk et al. (1977).

Column	Name	Value
1	year	year of data collection, 1974 or 1975
2	stratum	stratum membership, based on depth
3	catchnum	number of fish caught during trawl
4	catchwt	total weight (kg) of fish caught during trawl
5	numspp	number of species of fish caught during trawl
6	depth	depth of station (m)
7	temp	surface temperature (degrees C), missing = –99

otters.csv Data on number of holts (dens) in Shetland, U.K., used in Kruuk et al. (1989). Data courtesy of Hans Kruuk.

Column	Name	Value
1	section	section of coastline
2	habitat	type of habitat (stratum)
3	holts	number of holts (dens)

ozone.csv Hourly ozone readings (parts per billion, ppb) from a site in Monterey County, California, for 2018 and 2019. Source: `https://aqs.epa.gov/aqsweb/airdata/download_files.html#Raw`, accessed November 19, 2020. Missing values are denoted by −9.

Column	Name	Value
1	year	year of reading (2018 or 2019)
2	month	month of reading (1–12)
3	day	day of reading (1–31)
4	hr0	ozone reading (ppb) at 0:00 local time
5	hr1	ozone reading (ppb) at 1:00 local time
⋮	⋮	⋮
27	hr23	ozone reading (ppb) at 23:00 local time

pitcount.csv Fictional data from a fictional point-in-time (PIT) survey taken to estimate the number of persons experiencing homelessness.

Column	Name	Value
1	strat	Stratum number (from 1 to 8)
2	division	Geographic division, used to form strata
3	density	Expected density of persons experiencing homelessness (character variable, with values High or Low)
4	popsize	$= N_h$, the number of areas in the population for stratum h
5	sampsize	$= n_h$, the number of areas in the sample for stratum h
6	areawt	$= N_h/n_h$, the sampling weight for the area
7	y	Number of persons experiencing unsheltered homelessness found in the area during the PIT count

profresp.csv The data described in Zhang et al. (2020) were downloaded from `http://doi.org/10.3886/E109021V1` on January 22, 2020, from file `survey4.rds`. The data set `profresp.csv` contains selected variables from the set of 2,407 respondents who completed the survey and provided information on the demographic variables and the information needed to calculate "professional respondent" status. The full data set `survey4.rds` contains numerous additional questions about behavior that are not included here, as well as the data from the partially completed surveys. The website also contains data for three other online panel surveys. Because `profresp.csv` is a subset of the full data, statistics calculated from it may differ from those in Zhang et al. (2020).

Missing values are denoted by −9.

Column	Name	Value
1	prof_cat	Level of professionalism 1 = novice, 2 = average, 3 = professional
2	panelnum	Number of panels respondent has belonged to. A response between 1 and 6 means that the person has belonged to that number of panels; 7 means 7 or more.
3	survnum_cat	How many Internet surveys have you completed before this one? 1 = This is my first one, 2 = 1–5, 3 = 6–10, 4 = 11–15, 5 = 16–20, 6 = 21–30, 7 = More than 30
4	panelq1	Are you a member of any online survey panels besides this one? 1 = yes, 2 = no
5	panelq2	To how many other online panels do you belong?

profresp.csv (continued)

Column	Name	Value
		1 = None, 2 = 1 other panel, 3 = 2 others, 4 = 3 others, 5 = 4 others, 6 = 5 others, 7 = 6 others or more. This question has a missing value if *panelq1* = 2. If you want to estimate how many panels a respondent belongs to, create a new variable *numpanel* that equals *panelq2* if *panelq2* is not missing and equals 1 if *panelq1* = 2.
6	age4cat	Age category. 1 = 18 to 34, 2 = 35 to 49, 3 = 50 to 64, 4 = 65 and over
7	edu3cat	Education category. 1 = high school or less, 2 = some college or associates' degree, 3 = college graduate or higher
8	gender	Gender: 1 = male, 2 = female
9	non_white	1 = race is non-white, 0 = race is white
10	motive	Which best describes your main reason for joining on-line survey panels? 1 = I want my voice to be heard, 2 = Completing surveys is fun, 3 = To earn money, 4 = Other (Please specify)
11	freq_q1	During the PAST 12 MONTHS, how many times have you seen a doctor or other health care professional about your own health? Response is number between 0 and 999.
12	freq_q2	During the PAST MONTH, how many days have you felt you did not get enough rest or sleep?
13	freq_q3	During the PAST MONTH, how many times have you eaten in restaurants? Please include both full-service and fast food restaurants.
14	freq_q4	During the PAST MONTH, how many times have you shopped in a grocery store? If you shopped at more than one grocery store on a single trip, please count them separately.
15	freq_q5	During the PAST 2 YEARS, how many overnight trips have you taken?

profrespacs.csv Population estimates from the 2011 American Community Survey (ACS) for age/gender/education categories measured in `profresp.csv` (Zhang et al., 2020). Note that *age3cat* has 3 categories, while the age variable in `profresp.csv` has 4 categories.

Column	Name	Value
1	gender	Gender: 1 = male, 2 = female
2	age3cat	Age category. 1 = 18 to 34, 2 = 35 to 64, 3 = 65 and over
3	edu3cat	Education category. 1 = high school or less, 2 = some college or associates' degree, 3 = college graduate or higher
4	count	Population size from ACS for the gender/age/education level combination

radon.csv Radon readings for a stratified sample of 1003 homes in Minnesota. Source: Nolan and Speed (2000). The data were downloaded in April 2008 from an earlier version of the website now located at `www.stat.berkeley.edu/users/statlabs/labs.html`.

Column	Name	Value
1	countyname	County Name
2	countynum	County Number
3	sampsize	Sample size in county
4	popsize	Population size in county
5	radon	Radon concentration (picocuries per liter)

rectlength.csv Lengths of rectangles.

Column	Name	Value
1	rectangle	Rectangle number
2	length	Rectangle length

rnt.csv Page from a random number table. Open the `.csv` file in a text editor instead of a spreadsheet, because a spreadsheet strips off the leading zeroes. The columns have format z5.0 in the SAS file, and are character variables in the R file, so that leading zeroes are displayed in those formats.

Column	Name	Value
1	col1	Column of 5-digit random numbers
2	col2	Column of 5-digit random numbers
3	col3	Column of 5-digit random numbers
4	col4	Column of 5-digit random numbers
5	col5	Column of 5-digit random numbers
6	col6	Column of 5-digit random numbers

sample70.csv All possible simple random samples that can be generated from the population in Example 2.2 of SDA.

Column	Name	Value
1	sampnum	Sample number
2–5	u1–u4	Sampled units in $\mathcal{S}$
6–9	y1–y4	Values of y_i in sample $\mathcal{S}$
10	total	Estimated population total

santacruz.csv The number of seedlings in the sampled psus on Santa Cruz Island, California, in 1992 and 1994. Source: Peart (1994).

Column	Name	Value
1	tree	Tree number
2	seed92	Number of seedlings in 1992
3	seed94	Number of seedlings in 1994

schools.csv Math and reading test results from a two-stage cluster sample of tenth-grade students. An SRS of 10 schools was selected from the 75 schools in the population, and then 20 students were sampled from each school. These data are fictional but the summary statistics are consistent with those seen in educational studies.

Column	Name	Value
1	schoolid	School number (use as cluster variable)
2	gender	Gender of student (character variable, F = female, M = male)
3	math	Score on math test
4	reading	Score on reading test
5	mathlevel	Category level for math test score: 1 if $1 \leq$ math $<= 40$ 2 if $41 \leq$ math
6	readlevel	Category level for reading test score: 1 if $1 \leq$ read $<= 32$ 2 if $33 \leq$ read $<= 50$
7	Mi	Number of students in school, M_i
8	finalwt	Weight for student in sample

seals.csv Data on number of breathing holes found in sampled areas of Svalbard fjords, reconstructed from summary statistics given in Lydersen and Ryg (1991).

Column	Name	Value
1	zone	zone number for sampled area
2	holes	number of breathing holes Imjak found in area

shapespop.csv Population of black and gray squares and circles.

Column	Name	Value
1	ID	identification number for object
2	shape	shape of object (square or circle)
3	color	color of object (gray or black)
4	area	area of object (cm^2)
5	conv	= 1 if object can be reached through convenience sample, 0 otherwise

shorebirds.csv Two-phase sample of shorebird nests. These are artificial data constructed from summary statistics given in Bart and Earnst (2002).

Column	Name	Value
1	plot	Plot number
2	rapid	Rapid-method count of number of birds in plot
3	intense	Intensive-method count of number of nests in plot $= -9$ if the plot is not in the phase II sample

sp500.csv Companies in the S&P 500® Stock Market Index as of September 15, 2020. Source: Downloaded from `https://fknol.com/list/eps-sp-500-index-companies.php` on September 19, 2020.

Column	Name	Value
1	Company	Company name (character variable, length 37)
2	Symbol	Stock symbol (character variable, length 5)
3	MarketCap	Market capitalization, in billions of U.S. dollars
4	StockPrice	Price per share of stock
5	PE_Ratio	Price-to-earnings ratio
6	EPS	Earnings per share

spanish.csv Fictional cluster sample of introductory Spanish students.

Column	Name	Value
1	class	Class number
2	score	Score on vocabulary test (out of 100)
3	trip	= 1 if plan a trip to a Spanish-speaking country, 0 otherwise

srs30.csv An SRS of size 30 taken from an artificial population of size 100.

Column	Name	Value
1	y	Value of observation

ssc.csv SRS of 150 members of the Statistical Society of Canada, downloaded from `ssc.ca` in August, 2006.

Column	Name	Value
1	gender	m = male, f=female
2	occupation	a = academic, g = government, i = industry, n = not determined
3	ASA	= 1 if person is member of American Statistical Association, 0 otherwise

statepop.csv Data from an unequal-probability sample of 100 counties from the 1994 *County and City Data Book* (U.S. Census Bureau, 1994). The sample was selected with probability proportional to population.

Column	Name	Value
1	county	county name (character variable, length 14)
2	state	state name (character variable)
3	landarea	land area of county, 1990 (square miles)
4	popn	population of county, 1992
5	phys	number of physicians, 1990
6	farmpop	farm population, 1990
7	numfarm	number of farms, 1987
8	farmacre	number of acres devoted to farming, 1987
9	veterans	number of veterans, 1990
10	percviet	percent of veterans from Vietnam era, 1990
11	psii	ψ_i, probability of selection
12	wt	sampling weight, $= 1/(100\psi_i)$

statepps.csv Number of counties (or county equivalents; Alaska has boroughs, Louisiana has parishes, and some states have independent cities), population estimates for 2019, land area, and water area for the 50 states plus the District of Columbia. Total area for a state can be calculated by summing land area and water area. Source: Population estimates are from U.S. Census Bureau (2019). Land and water areas are from U.S. Census Bureau (2012).

Column	Name	Value
1	state	state name (character variable, length 20)
2	counties	number of counties or county equivalents
3	pop2019	population of state, 2019
4	landarea	land area of state (square kilometers)
5	waterarea	water area of state (square kilometers)

swedishlcs.csv Data on call attempts from the Swedish Survey of Living Conditions. Source: Lundquist and Särndal (2013).

Column	Name	Value
1	attempt	call attempt number
2	resprate	response rate at call attempt (percent)
3	benefits	relative bias for variable *benefits*
4	income	relative bias for variable *income*
5	employed	relative bias for variable *employed*
6	note	Character variable, length 25: notes about data collection

The variable *attempt* takes on values 1–25 for the initial fieldwork period. Values 31–40 denote the follow-up period, and value 45 gives the final estimates. The gaps in the attempt variable allow one to see the separation of the periods on the graph.

syc.csv Selected variables from the Survey of Youth in Custody (Beck et al., 1988). Source: U.S. Department of Justice (1989). Strata 6–16 each contain one facility; the psus in those strata are residents. In strata 1–5, the psus are facilities. The number of facilities in the population (N_h) for those five strata are: $N_1 = 99$, $N_2 = 39$, $N_3 = 30$, $N_4 = 13$, $N_5 = 14$. Eleven facilities are sampled from stratum 1, and seven facilities are sampled from each of strata 2–5.

The table gives missing value codes for individual variables in the `.csv` file (these codes are the same as in the original data source, but have been changed to the appropriate missing value codes for the respective software packages in the SAS and R data files).

Column	Name	Value
1	stratum	stratum number
2	psu	psu number, = facility number for residents in strata 1–5 and person number for residents in strata 6–16
3	facility	facility number
4	facsize	number of eligible residents in psu
5	finalwt	final weight
6	randgrp	random group number
7	age	age of resident (99=missing)

syc.csv (continued)

Column	Name	Value
8	race	race of resident 1 = white; 2 = Black; 3 = Asian/Pacific Islander; 4 = American Indian, Alaska Native; 5 = Other; 9 = Missing
9	ethnicty	1 = Hispanic, 2 = not Hispanic, 9=missing
10	educ	highest grade attended before sent to correctional institution 0 = Never attended school; 1–12 = highest grade attended; 13 = GED; 14 = Other
11	gender	1 = male, 2 = female
12	livewith	Who did you live with most of the time you were growing up? 1 = Mother only, 2 = Father only 3 = Both mother and father, 4 = Grandparents, 5 = Other relatives, 6 = Friends, 7 = Foster home, 8 = Agency or institution, 9 = Someone else, 99 = Blank
13	famtime	Has anyone in your family, such as your mother, father, brother, sister, ever served time in jail or prison? 1 = Yes, 2 = No, 7 = Don't know, 9 = Blank
14	crimtype	most serious crime in current offense 1 = violent (e.g., murder, rape, robbery, assault) 2 = property (e.g., burglary, larceny, arson, fraud, motor vehicle theft) 3 = drug (drug possession or trafficking) 4 = public order (weapons violation, perjury, failure to appear in court) 5 = juvenile status offense (truancy, running away, incorrigible behavior) 9 = missing
15	everviol	ever put on probation or sent to correctional inst for violent offense: 1 = yes, 0 = no
16	numarr	number of times arrested (99=missing)
17	probtn	number of times on probation (99=missing)
18	corrinst	number of times previously committed to correctional institution (99=missing)
19	evertime	Prior to being sent here did you ever serve time in a correctional institution? 1 = yes, 2 = no, 9 = missing
20	prviol	=1 if previously arrested for violent offense, 0 otherwise
21	prprop	=1 if previously arrested for property offense, 0 otherwise
22	prdrug	=1 if previously arrested for drug offense, 0 otherwise
23	prpub	=1 if previously arrested for public order offense, 0 otherwise
24	prjuv	=1 if previously arrested for juvenile status offense, 0 otherwise
25	agefirst	age first arrested (99=missing)
26	usewepn	Did you use a weapon ... for this incident? 1 = Yes, 2 = No, 9 = Blank
27	alcuse	Did you drink alcohol at all during the year before being sent here this time? 1 = Yes; 2 = No, didn't drink during year before; 3 = No, don't drink at all, 9 = missing
28	everdrug	Ever used illegal drugs; 0=no, 1=yes, 9=missing

teachers.csv Selected variables from a study on elementary school teacher workload in Maricopa County, Arizona. Data courtesy of Rita Gnap (Gnap, 1995). The psu sizes are given in file `teachmi.csv`. The large stratum had 245 schools; the small/medium stratum had 66 schools. Missing values are coded as −9.

Column	Name	Value
1	dist	school district size. Character variable: large or med/small
2	school	school identifier
3	hrwork	number of hours required to work at school per week
4	size	class size
5	preprmin	minutes spent per week in school on preparation
6	assist	minutes per week that a teacher's aide works with the teacher in the classroom

teachmi.csv Cluster sizes for data in `teachers.csv`.

Column	Name	Value
1	dist	School district size: large or med/small
2	school	school identifier
3	popteach	number of teachers in that school
4	ssteach	number of surveys returned from that school

teachnr.csv Data from a follow-up study of nonrespondents from Gnap (1995).

Column	Name	Value
1	hrwork	number of hours required to work at school per week
2	size	class size
3	preprmin	minutes spent per week in school on preparation
4	assist	minutes per week that a teacher's aide works with the teacher in the classroom

uneqvar.csv Artificial data used in exercises of Chapter 11.

Column	Name	Value
1	x	x
2	y	y

vietnam.csv Vietnam-service data from Stockford and Page (1984).

Column	Name	Value
1	apc	APC stratum. Character variable with options "Yes," "No," "NotAvail"
2	p2sample	Indicator variable for phase II sample, = 1 if in phase II sample, 0 otherwise
3	vietnam	= 1 if service in Vietnam, = 0 if service not in Vietnam, = −9 if not in phase II sample

vietnam.csv (continued)

Column	Name	Value
4	phase1wt	weight for phase I sample
5	phase2wt	conditional weight for phase II sample, calculated as (phase I sample size in stratum) / (phase II sample size in stratum). phase2wt = −9 for observations not in phase 2 sample.
6	finalwt	final weight for phase II sample, calculated as phase1wt*phase2wt (= −9 for observations not in phase II sample)
7	p1apcsize	number of observations in the observation's APC stratum that are in the phase I sample (n_h)
8	p2apcsize	number of observations in the observation's APC stratum that are in the phase II sample (m_h)

vius.csv Selected variables from the 2002 U.S. Vehicle Inventory and Use Survey (VIUS). Source: U.S. Census Bureau (2006). The data were downloaded from `www.census.gov/svsd/www/vius` in May, 2006. The website from which the data were downloaded no longer exists, and online information about VIUS may now be found at `https://www.bts.gov/vius`, which provides a link to the archived 2002 data. The missing value of *state* for records with *adm_state* = 42 was recoded to "PA," the state that has code 42. This data set has 98,682 records, which may be too large for some software packages to handle; the file `viusca.csv` is a smaller data set, with the same columns described below, containing only vehicles from California. The variable descriptions below are taken from the VIUS Data Dictionary.

Missing values are coded as −99. For some variables, the value is missing because the question is not applicable or the vehicle is not in use; see the individual variable descriptions.

Note that a new VIUS is planned for 2022, with data to be released in 2023; see `https://www.bts.gov/vius`.

Column	Name	Value
1	stratum	stratum number (contains all 255 strata)
2	adm_state	state number
3	state	state name
4	trucktype	type of truck, used in stratification 1. pickups 2. minivans, other light vans, and sport utility vehicles 3. light single-unit trucks with gross vehicle weight less than 26,000 pounds 4. heavy single-unit trucks with gross vehicle weight greater than or equal to 26,000 pounds 5. truck-tractors
5	tabtrucks	column of sampling weights
6	bodytype	body type of vehicle 01. Pickup 02. Minivan 03. Light van other than minivan 04. Sport utility

vius.csv (continued)

Column	Name	Value
		05. Armored
		06. Beverage
		07. Concrete mixer
		08. Concrete pumper
		09. Crane
		10. Curtainside
		11. Dump
		12. Flatbed, stake, platform, etc.
		13. Low boy
		14. Pole, logging, pulpwood, or pipe
		15. Service, utility
		16. Service, other
		17. Street sweeper
		18. Tank, dry bulk
		19. Tank, liquids or gases
		20. Tow/Wrecker
		21. Trash, garbage, or recycling
		22. Vacuum
		23. Van, basic enclosed
		24. Van, insulated non-refrigerated
		25. Van, insulated refrigerated
		26. Van, open top
		27. Van, step, walk-in, or multistop
		28. Van, other
		99. Other not elsewhere classified
7	adm_modelyear	model year
		01. 2003, 2002
		02. 2001
		03. 2000
		04. 1999
		05. 1998
		06. 1997
		07. 1996
		08. 1995
		09. 1994
		10. 1993
		11. 1992
		12. 1991
		13. 1990
		14. 1989
		15. 1988
		16. 1987
		17. Pre-1987
8	vius_gvw	Gross vehicle weight based on average reported weight
		01. Less than 6,001 lbs.
		02. 6,001 to 8,500 lbs.
		03. 8,501 to 10,000 lbs.
		04. 10,001 to 14,000 lbs.

vius.csv (continued)

Column	Name	Value
		05. 14,001 to 16,000 lbs.
		06. 16,001 to 19,500 lbs.
		07. 19,501 to 26,000 lbs.
		08. 26,001 to 33,000 lbs.
		09. 33,001 to 40,000 lbs.
		10. 40,001 to 50,000 lbs.
		11. 50,001 to 60,000 lbs.
		12. 60,001 to 80,000 lbs.
		13. 80,001 to 100,000 lbs.
		14. 100,001 to 130,000 lbs.
		15. 130,001 lbs. or more
9	miles_annl	Number of Miles Driven During 2002
10	miles_life	Number of Miles Driven Since Manufactured
11	mpg	Miles Per Gallon averaged during 2002. Range from 0.3 to 35. −99 denotes not reported or not applicable.
12	opclass	Operator Classification With Highest Percent
		1. Private
		2. Motor carrier
		3. Owner operator
		4. Rental
		5. Personal transportation
		6. Not applicable (Vehicle not in use)
13	opclass_mtr	Percent of Miles Driven as a Motor Carrier. −99 denotes vehicle not in use
14	opclass_own	Percent of Miles Driven as an Owner Operator. −99 denotes vehicle not in use
15	opclass_psl	Percent of Miles Driven for Personal Transportation. −99 denotes vehicle not in use
16	opclass_pvt	Percent of Miles Driven as Private (Carry Own Goods or Internal Company Business Only). −99 denotes vehicle not in use
17	opclass_rnt	Percent of Miles Driven as Rental. −99 denotes vehicle not in use
18	transmssn	Type of Transmission
		1. Automatic
		2. Manual
		3. Semi-Automated Manual
		4. Automated Manual
19	trip_primary	Primary Range of Operation
		1. Off-the-road
		2. Less than 50 miles
		3. 51 to 100 miles
		4. 101 to 200 miles
		5. 201 to 500 miles
		6. 501 miles or more
		7. Not reported
		8. Not applicable (Vehicle not in use)
20	trip0_50	Percent of Annual Miles Accounted for with Trips

vius.csv (continued)		
Column	**Name**	**Value**
		50 Miles or Less from the Home Base
21	trip051_100	Percent of Annual Miles Accounted for with Trips 51 to 100 Miles from the Home Base
22	trip101_200	Percent of Annual Miles Accounted for with Trips 101 to 200 Miles from the Home Base
23	trip201_500	Percent of Annual Miles Accounted for with Trips 201 to 500 Miles from the Home Base
24	trip500more	Percent of Annual Miles Accounted for with Trips 501 or More Miles from Home Base
25	adm_make	Make of vehicle
		01. Chevrolet
		02. Chrysler
		03. Dodge
		04. Ford
		05. Freightliner
		06. GMC
		07. Honda
		08. International
		09. Isuzu
		10. Jeep
		11. Kenworth
		12. Mack
		13. Mazda
		14. Mitsubishi
		15. Nissan
		16. Peterbilt
		17. Plymouth
		18. Toyota
		19. Volvo
		20. White
		21. Western Star
		22. White GMC
		23. Other (domestic)
		24. Other (foreign)
26	business	Business in which vehicle was most often used during 2002
		01. For-hire transportation or warehousing
		02. Vehicle leasing or rental
		03. Agriculture, forestry, fishing, or hunting
		04. Mining
		05. Utilities
		06. Construction
		07. Manufacturing
		08. Wholesale trade
		09. Retail trade
		10. Information services
		11. Waste management, landscaping, or administrative/support services

vius.csv (continued)

Column	Name	Value
		12. Arts, entertainment, or recreation services
		13. Accommodation or food services
		14. Other services
		−99. Not reported or not applicable

winter.csv Selected variables from the Arizona State University Winter Closure Survey, taken in January 1995 (provided courtesy of the ASU Office of University Evaluation). This survey was taken to investigate the attitudes and opinions of university employees towards the closing of the university (for budgetary reasons) between December 25 and January 1. For the yes/no questions, the responses are coded as 1 = No, 2 = Yes. The variables *treatsta* and *treatme* are coded as 1=strongly agree, 2=agree, 3=undecided, 4=disagree, 5=strongly disagree. The variables *process* and *satbreak* are coded as 1=very satisfied, 2=satisfied, 3=undecided, 4=dissatisfied, 5=very dissatisfied. Variables *ownsupp* through *offclose* are coded 1 if the person checked that the statement applied to him/her, and 2 if the statement was not checked.

Missing values are coded as 9.

Column	Name	Value
1	class	stratum number 1 = faculty ; 2 = classified staff; 3 = administrative staff; 4 = academic professional
2	yearasu	number of years worked at ASU 1 = 1–2 years; 2 = 3–4 years; 3 = 5–9 years; 4 = 10–14 years; 5 = 15 or more years
3	vacation	In the past, have you *usually* taken vacation days the entire period between December 25 and January 1?
4	work	Did you work on campus during Winter Break Closure?
5	havediff	Did the Winter Break Closure cause you any difficulty/concerns?
6	negaeffe	Did the Winter Break Closure *negatively* affect your work productivity?
7	ownsupp	I was unable to obtain staff support in my department/office
8	othersup	I was unable to obtain staff support in other departments/offices
9	utility	I was unable to access computers, copy machine, etc. in my department/office
10	environ	I was unable to endure environmental conditions, e.g., not properly climatized
11	uniserve	I was unable to access university services necessary to my work
12	workelse	I was unable to work on my assignments because I work in another department/office
13	offclose	I was unable to work on my assignments because my office was closed
14	treatsta	Compared to other departments/offices, I feel staff in my department/office were treated fairly

winter.csv (continued)

Column	Name	Value
15	treatme	Compared to other people working in my department/office, I feel I was treated fairly
16	process	How satisfied are you with the process used to inform staff about Winter Break Closure?
17	satbreak	How satisfied are you with the fact that ASU had a Winter Break Closure this year?
18	breakaga	Would you want to have Winter Break Closure again?

wtshare.csv Hypothetical sample of size 100, with indirect sampling. The data set has multiple records for adults with more than one child; if adult 254 has 3 children, adult 254 is listed 3 times in the data set. Note that to obtain L_k, you need to take *numadult* +1.

Column	Name	Value
1	id	Identification number of adult in sample
2	child	= 1 if record is for a child, 0 if adult has no children
3	preschool	= 1 if child is in preschool, 0 otherwise
4	numadult	number of other adults in population who link to that child

B

Jackknife Macros

This appendix presents two macros for using jackknife methods with survey data analyses. The first, in Section B.1, uses jackknife replicate weights and the CATMOD procedure to calculate standard errors for loglinear models. It is easily modified to calculate estimates using replicate weights for other non-survey procedures that allow estimates to be computed using sampling weights. SAS Institute Inc. (2015) presents a similar macro for fitting Poisson regression models.

The macro in Section B.2 computes jackknife replicate weights for a simple two-phase sample in which a simple random sample (SRS) is selected at phase I and a stratified random sample is selected in phase II, where the strata are defined by information collected in the phase I sample. This macro was used in Section 12.1.

B.1 Using Replicate Weights with Non-Survey Procedures

As mentioned in Section 10.3, replicate weight methods can be used to calculated standard errors for statistics calculated using non-survey SAS software procedures that allow estimates to be calculated with weights.

This is done by calculating the value of the statistic of interest using the final weight and then using each column of replicate weights, then using the appropriate variance formula to calculate the variance from the replicate statistics. The same code is executed for each replicate weight, and this is best done in a macro.

The macro in Code B.1 and in file `jkcatmod.sas` calculates loglinear model coefficients for the final weight and each set of replicate weights using the CATMOD procedure for Example 10.9 of SDA. This macro was used in Section 10.3 to calculate the coefficients for each replicate. This example uses jackknife replicate weights, but statistics are calculated the same way for any type of replicate weights.

Code B.1. A macro to calculate standard errors for loglinear model coefficients via jackknife (`jkcatmod.sas`).

```
%macro jkcatmod(jkdata=,replicates=,wtvar=,factors=,llmodel=);

/* Arguments
   jkdata      name of data set containing data and JK replicate weights.
               These must be named RepWt_1 ... RepWt_&replicates.
   replicates number of replicate weights
   wtvar       name of weight variable in full data
   factors     response variables
   llmodel     model to be fit in LOGLIN statement
```

DOI: 10.1201/9781003160366-B

```
   Data sets created
   estfull     estimates from full data with weight wtvar
   estjk       estimates from jackknife iterations
   convfull    convergence status from full model with weight wtvar
   convjk      convergence status from jackknife iterations
*/

/* Turn off output while macro executes */

OPTIONS NONOTES;
ODS SELECT NONE;

/* Fit full model */
   PROC CATMOD DATA=&jkdata;
      WEIGHT &wtvar;
      MODEL &factors = _RESPONSE_;
      LOGLIN &llmodel;
      ODS OUTPUT ESTIMATES = estfull convergencestatus = convfull;
         TITLE 'Full model';
   RUN;
   DATA estfull (keep = Parameter ClassValue Estimatefull);
      SET estfull;
      estimatefull = estimate;

/* Fit the model with each replicate weight */
%DO i=1 %TO &replicates;
    PROC CATMOD DATA=&jkdata   ;
      WEIGHT RepWt_&i;
      MODEL &factors = _RESPONSE_;
      LOGLIN &llmodel;
      ODS OUTPUT ESTIMATES = estrep convergencestatus = convrep;
      TITLE "Estimate from jk replicate &i";
    RUN;

    DATA estrep (keep = Parameter ClassValue Estimate replicate);
      SET estrep;
      replicate = &i;
    DATA convrep;
      SET convrep;
      replicate = &i;
        RUN;

    %IF &i = 1 %THEN %DO;
      DATA estjk;
         SET estrep;
      DATA convjk;
         SET convrep;
      RUN;
    %END;
    %ELSE %DO;
      PROC APPEND BASE=estjk DATA=estrep;
      PROC APPEND BASE=convjk DATA=convrep;
      RUN;
    %END;
%END;
```

```
ODS SELECT ALL;
OPTIONS NOTES;
%mend jkcatmod;
```

To run the macro with an independence model on data set *sycjkwts* containing the data values and replicate weights, type:

```
%jkcatmod(jkdata=sycjkwts, replicates=116, wtvar=finalwt, factors=ageclass*everviol
    *famtime, llmodel=ageclass everviol famtime);
```

This executes the code in the macro and stores the estimates in the data sets *estfull* and *estjk*. At the end of each jackknife iteration in the %DO loop, the parameter estimates from the replicate weight variable used in that iteration are appended to the parameter estimates in data set *estjk*.

Loglinear model coefficients are estimated using an iterative process, so for this analysis, data sets *convfull* and *convjk* contain the convergence status for the model fit with the final weights and for the models fit with each set of replicate weights. Convergence status for the full data set and each replicate should be checked before proceeding with the analysis.

Variable *estimatefull* in data set *estfull* contains the parameter estimates from the full data set, and variable *estimate* in data set *estjk* contains the parameter estimates from the replicate weights. These values, and the jackknife coefficients, all need to be in the same data set to calculate the standard errors. The following code merges the data sets.

```
/* Merge the data set containing jackknife replicates with the jackknife
    coefficients */

PROC SORT DATA=estjk;
  BY replicate;
PROC SORT DATA=sycjkcoef;
  BY replicate;
DATA estjk1;
  MERGE estjk sycjkcoef;
  BY replicate;

/* Merge again with the data set containing estimates from the full data */

PROC SORT DATA=estfull;
  BY parameter classvalue;
PROC SORT DATA=estjk1;
  BY parameter classvalue;

DATA calcjk;
  MERGE estfull estjk1;
  BY parameter classvalue;
  jkvarterm = (estimate - estimatefull)**2 * JKCoefficient;
RUN;
```

In the merged data set *calcjk*, variable *jkvarterm* contains $c_r(\hat{\theta}_r - \hat{\theta})^2$ from Equation (9.1) for each parameter and for each replicate r. The estimated variance for the parameter is then the sum of these values.

```
PROC MEANS DATA=calcjk SUM;
  BY parameter classvalue;
  var jkvarterm;
  OUTPUT OUT = jkvar SUM = jkvariance;
RUN;
```

Variable *jkvariance* contains the jackknife variance for each estimated parameter from the loglinear model.

This example calculates jackknife variances for point estimators calculated using the CATMOD procedure. The method is similar for other procedures and other types of replicate weights. If desired, the last steps of merging the data sets and calculating estimates can also be implemented in a macro.

Warning: Not all WEIGHT statements are the same. For the loglinear model, the WEIGHT statement in the CATMOD procedure can be used to calculate estimates with the survey weights. The CATMOD procedure uses the WEIGHT variable to denote frequency counts, which can be thought of as the number of population units represented by a unit in the sample data. But some procedures in SAS software use the WEIGHT statement in a different way—check to see what the WEIGHT variable represents before using the procedure to calculate estimates from a complex survey.

For a procedure that calculates maximum likelihood estimates, check in the Details section of the documentation that the weighted log-likelihood function used for estimating a parameter θ is of the form

$$\hat{\ell}(\theta) = \sum_{i \in \mathcal{S}} w_i \ln f(y_i, \theta), \tag{B.1}$$

where f is the density function for the assumed model. This log-likelihood function estimates

$$\ell(\theta) = \sum_{i=1}^{N} \ln f(y_i, \theta),$$

the log-likelihood that would be obtained if the entire population were known. In (B.1), each observation in the sample $\mathcal{S}$ is weighted by w_i. Maximizing (B.1) gives a value $\hat{\theta}$ that estimates the value of θ that would maximize the log-likelihood for the entire population, $\ell(\theta)$; $\hat{\theta}$ is a pseudo-maximum-likelihood estimate for the model (Binder, 1983).

For pseudo-maximum-likelihood estimation to work, the procedure must treat the WEIGHT variable as a frequency count and allow non-integer values. Some procedures treat the WEIGHT variable in a different way. In the GENMOD procedure, for example, which produces estimates for a variety of generalized linear models, the WEIGHT statement refers to the exponential family dispersion parameter weight for each observation. For Poisson regression, the log-likelihood function with the weights has the form in (B.1). Thus, the Poisson regression parameters calculated by the GENMOD procedure with the WEIGHT variable estimate the maximum likelihood parameter estimates that would be obtained if the model were fit to the entire population.

But for negative binomial regression, the dispersion weight does not factor out of the log-likelihood. For negative binomial regression, the log-likelihood function $\hat{\ell}(\theta)$ maximized by the GENMOD procedure, when the WEIGHT statement is used, is not of the form in (B.1), and thus the procedure will not give the pseudo-maximum-likelihood estimates of the model parameters. Another method must be used to estimate model parameters for negative binomial regression (SAS Institute Inc., 2015).

B.2 Jackknife for Two-Phase Sampling

The macro given below and in file `JK2phase.sas` computes jackknife replicate weights and coefficients for a two-phase sample, using the method in Kim et al. (2006), when:

- Phase I is a simple random sample of size $n = numreps$.
- Phase II is a stratified random sample with m_h observations subsampled from the n_h observations in stratum h of Phase I.

It is assumed that the stratum membership is known for everyone in the Phase I sample and there are no missing values for y (this is true for most of the exercises in Chapter 12 of SDA). The macro can be modified to accommodate files with missing data.

The input data set *indata* must contain records for each phase I unit (including the units not sampled in phase II) for the replicate weights to be created correctly.

Macro *jk2phase* defines the phase II replicate weights using the phase I replicate weights (created for an SRS in the SURVEYMEANS procedure), the phase I stratum membership, and the phase II sample membership. It produces data set *p1jkwt*, containing the jackknife weights for the Phase I sample, *p2jkwt*, containing the jackknife weights for the Phase II sample (this data set contains only the observations that are in the Phase II sample), and data set *jkcoefs* containing the jackknife coefficients used for both sets of replicate weights. The replicate weights have variable names *RepWt_1, ..., RepWt_n* in each data set *p1jkwt* and *p2jkwt*; if you are performing analyses using information from both phases, rename the replicate weight variables before merging the data sets.

For a two-phase sample, the number of jackknife replicate weights equals the phase-I sample size, n. If n is large, the resulting data sets with the jackknife weights will be large (each will contain n columns of replicate weights), and subsequent computations with the survey analysis procedures may be slow. If n is very large, you may want to use a different variance estimation method.

Code B.2. A macro to calculate two-phase jackknife replicate weights (`JK2phase.sas`).

```
%macro jk2phase(indata=,p1jkwt=,p2jkwt=,jkcoefs=,numreps=,strat=,nh=,mh=,p1weight=,
    p2sample=);

/* Creates jackknife replicate weights for a two-phase survey where
     Phase I is an SRS
     Phase II is a stratified random sample,
        m_h units randomly sampled from the n_h Phase-I units in stratum h
     User should perform missing value checks before using macro */

/* Note: This macro creates a replicate weight for each observation in the phase I
   sample. If n is large, this will result in a large data set and slow
   computations. */

/* Arguments
   indata      name of input data set
               Must contain records for all phase I units
   p1jkwt      name of output data set to contain the replicate weights
                  for phase I sample
   p2jkwt      name of output data set to contain the replicate weights
```

```
                  for phase II sample
   jkcoefs      name of output data set to contain the jackknife coefficients
                  for both phases
   numreps      number of phase I observations (=n)
   strat        name of variable in indata containing stratum membership
   nh           name of variable in indata containing phase I sample size
                  in stratum h (n_h)
   mh           name of variable in indata containing phase II sample size
                  in stratum h (m_h)
   p1weight     name of variable in indata containing phase I weight (=N/n)
   p2sample     name of variable in indata indicating membership in phase II sample,
                  should equal 1 if record in phase II sample
*/

/* Turn off output while macro executes */
OPTIONS NONOTES;
ODS SELECT NONE;

/* Create phase 1 jackknife weights and coefficients */

PROC SURVEYMEANS DATA = &indata PLOTS=NONE
     VARMETHOD = JACKKNIFE (OUTWEIGHTS = &p1jkwt OUTJKCOEFS = &jkcoefs);
  WEIGHT &p1weight;
  /* Can list any variable name in VAR statement since using procedure only to
   create weights */
  VAR &p2sample;
RUN;

/* Create stratum and phase II indicator variables to correspond to replicate
   weight columns */

PROC TRANSPOSE DATA = &indata OUT = str_tr PREFIX = str_;
  VAR &strat;

PROC TRANSPOSE DATA = &indata OUT = p2ind_tr PREFIX = p2_;
  VAR &p2sample;

DATA &p2jkwt (DROP = str_1 - str_&numreps p2_1 - p2_&numreps i _NAME_);
  SET &p1jkwt;
  IF _N_ = 1 THEN DO;
    SET str_tr;
        SET p2ind_tr;
  END;
  ARRAY RepWt {&numreps} RepWt_1 - RepWt_&numreps;
  /* Column strcol{i} contains stratum membership of ith observation */
  ARRAY strcol {&numreps} str_1 - str_&numreps;
  /* Column p2col{i} contains phase II membership status of ith observation */
  ARRAY p2col {&numreps} p2_1 - p2_&numreps;

  IF &p2sample = 1;  /* keep only phase II sample units */

  /* Calculate phase II weight using phase I weight, stratum membership, and phase
    II membership */
  DO i = 1 TO &numreps;
    IF &strat ne strcol{i} THEN RepWt{i} = RepWt{i}*&nh/&mh;
```

```
      ELSE IF &strat = strcol{i} AND p2col{i} = 0 THEN
              RepWt{i} = RepWt{i}*(&nh - 1)/(&mh);
      ELSE IF &strat = strcol{i} AND p2col{i} = 1 THEN
              RepWt{i} = RepWt{i}*(&nh - 1)/(&mh - 1);
   END;
RUN;

ODS SELECT ALL;
OPTIONS NOTES;
%mend jk2phase;  /* end macro jk2phase */
```

Fuller (2009, Section 4.4) suggests a modification of the phase II replicate weights that can result in more accurate jackknife variance estimates when the phase II sample sizes are small.

Bibliography

Allison, P. D. (2008). Convergence failures in logistic regression. In *Proceedings of the SAS® Global Forum 2008*, Number 360–2008. Cary, NC: SAS Institute, Inc. `https://support.sas.com/resources/papers/proceedings/pdfs/sgf2008/360-2008.pdf` (accessed December 9, 2020).

Allison, P. D. (2012). *Logistic Regression using SAS®: Theory and Application.* Cary, NC: SAS Institute, Inc.

An, T. (2020). A SAS® macro for calibration of survey weights. In *Proceedings of the SAS® Global Forum 2020*, Number 4284–2020. Cary, NC: SAS Institute Inc. `https://www.sas.com/content/dam/SAS/support/en/sas-global-forum-proceedings/2020/4284-2020.pdf` (accessed November 9, 2020).

Arnold, T. W. (1991). Intraclutch variation in egg size of American coots. *The Condor 93*, 19–27.

Bart, J. and S. Earnst (2002). Double-sampling to estimate density and population trends in birds. *The Auk 119*, 36–45.

Beck, A. J., S. A. Kline, and L. A. Greenfeld (1988). Survey of Youth in Custody. Technical Report NCJ-113365, Bureau of Justice Statistics, Washington, DC.

Berglund, P. A. (2015). Multiple imputation using the fully conditional specification method: A comparison of SAS®, Stata, IVEware, and R. In *Proceedings of the SAS® Global Forum 2015*, Number 2081–2015. Cary, NC: SAS Institute Inc. `https://support.sas.com/resources/papers/proceedings15/2081-2015.pdf` (accessed November 3, 2020).

Berglund, P. A. and S. G. Heeringa (2014). *Multiple Imputation of Missing Data using SAS®*. Cary, NC: SAS Institute, Inc.

Binder, D. A. (1983). On the variances of asymptotically normal estimators from complex surveys. *International Statistical Review 51*, 279–292.

Brewer, K. R. W. (1963). Ratio estimation and finite populations: Some results deducible from the assumption of an underlying stochastic process. *The Australian Journal of Statistics 5*(3), 93–105.

Brewer, K. R. W. (1975). A simple procedure for sampling πpswor. *The Australian Journal of Statistics 17*, 166–172.

Buskirk, T. D. (2011). Estimating design effects for means, proportions and totals from complex sample survey data using SAS® PROC SURVEYMEANS. In *Proceedings of the Midwest SAS® User Group Conference.* Cary, NC: SAS Institute, Inc. `https://mwsug.org/proceedings/2011/stats/MWSUG-2011-SA11.pdf` (accessed November 20, 2020).

Centers for Disease Control and Prevention (2017). NHANES Questionnaires, Datasets, and Related Documentation. `https://wwwn.cdc.gov/nchs/nhanes/` (accessed August 15, 2020).

Chauvet, G. and Y. Tillé (2006). A fast algorithm for balanced sampling. *Computational Statistics 21*(1), 53–62.

Chauvet, G. and Y. Tillé (2007). Application of fast SAS macros for balancing samples to the selection of addresses. *Case Studies in Business, Industry and Government Statistics (CSBIGS) 1*(2), 173–182.

Cormack, R. M. (1992). Interval estimation for mark-recapture studies of closed populations. *Biometrics 48*, 567–576.

Diaz-Ramirez, L. G., B. Jing, K. E. Covinsky, and W. J. Boscardin (2020). Mixed-effects models and complex survey data with the GLIMMIX procedure. In *Proceedings of the SAS® Global Forum 2020*, Number 4937–2020. Cary, NC: SAS Institute Inc. `https://www.sas.com/content/dam/SAS/support/en/sas-global-forum-proceedings/2020/4937-2020.pdf` (accessed January 21, 2021).

Eilers, P. H. C. and B. D. Marx (2021). *Practical Smoothing: The Joys of P-splines.* Cambridge: Cambridge University Press.

Fienberg, S. E. and A. Rinaldo (2007). Three centuries of categorical data analysis: Log-linear models and maximum likelihood estimation. *Journal of Statistical Planning and Inference 137*(11), 3430–3445.

Forman, S. L. (2004). Baseball-reference.com—Major league statistics and information. `www.baseball-reference.com` (accessed November 2004).

Fuller, W. A. (2009). *Sampling Statistics.* Hoboken, NJ: Wiley.

Gini, C. and L. Galvani (1929). Di una applicazione del metodo rappresentativo all'ultimo censimento italiano della popolazione. *Annali di Statistica 6*(4), 1–105.

Gnap, R. (1995). *Teacher Load in Arizona Elementary School Districts in Maricopa County.* Ph.D. dissertation. Tempe, AZ: Arizona State University.

Hand, D. J., F. Daly, A. D. Lunn, K. J. McConway, and E. Ostrowski (1994). *A Handbook of Small Data Sets.* London: Chapman and Hall.

Hanurav, T. V. (1967). Optimum utilization of auxiliary information: πps sampling of two units from a stratum. *Journal of the Royal Statistical Society, Series B 29*, 374–391.

Hayat, M. and T. Knapp (2017). Randomness and inference in medical and public health research. *Journal of the Georgia Public Health Association 7*(1), 7–11.

Haziza, D. (2009). Imputation and inference in the presence of missing data. In D. Pfeffermann and C. R. Rao (Eds.), *Sample Surveys: Design, Methods, and Applications. Handbook of Statistics, Volume 29A*, pp. 215–246. Amsterdam: North-Holland.

Heck, P. R., D. J. Simons, and C. F. Chabris (2018). 65% of Americans believe they are above average in intelligence: Results of two nationally representative surveys. *PloS One 13*(7), 1–11.

Hidiroglou, M. A., J.-F. Beaumont, and W. Yung (2019). Development of a small area estimation system at Statistics Canada. *Survey Methodology 45*(1), 101–126.

Hsu, J. (1996). *Multiple Comparisons: Theory and Methods.* Boca Raton, FL: CRC Press.

Hsu, J. C. and M. Peruggia (1994). Graphical representations of Tukey's multiple comparison method. *Journal of Computational and Graphical Statistics 3*(2), 143–161.

Hyndman, R. J. and Y. Fan (1996). Sample quantiles in statistical packages. *The American Statistician 50*(4), 361–365.

Izrael, D., M. P. Battaglia, A. A. Battaglia, and S. W. Ball (2017). You do not have to step on the same rake: SAS® raking macro generation IV. In *Proceedings of the SAS® 2017 Global Forum*, Number 0470–2017. Cary, NC: SAS Institute, Inc. https://support.sas.com/resources/papers/proceedings17/0470-2017-poster.pdf (accessed November 3, 2020).

Izrael, D., D. C. Hoaglin, and M. P. Battaglia (2000). A SAS® macro for balancing a weighted sample. In *Proceedings of the Twenty-Fifth Annual SAS® Users Group International Conference.* Cary, NC: SAS Institute, Inc. https://support.sas.com/resources/papers/proceedings/proceedings/sugi25/25/st/25p258.pdf (accessed November 3, 2020).

Jacoby, J. and A. H. Handlin (1991). Non-probability sampling designs for litigation surveys. *Trademark Reporter 81*, 169–179.

Kim, J. K. and W. Fuller (2004). Fractional hot deck imputation. *Biometrika 91*(3), 559–578.

Kim, J. K., A. Navarro, and W. A. Fuller (2006). Replication variance estimation for two-phase stratified sampling. *Journal of the American Statistical Association 101*, 312–320.

Koenker, R. (2005). *Quantile Regression.* Cambridge: Cambridge University Press.

Kruuk, H., A. Moorhouse, J. W. H. Conroy, L. Durbin, and S. Frears (1989). An estimate of numbers and habitat preferences of otters *Lutra lutra* in Shetland, UK. *Biological Conservation 49*, 241–254.

Lewis, T. H. (2012). Weighting adjustment methods for nonresponse in surveys. In *Proceedings of the Conference of Western Users of SAS® Software.* Cary, NC: SAS Institute, Inc. https://www.wuss.org/future-and-past-conferences-and-proceedings/ (accessed November 3, 2020).

Lewis, T. H. (2015). Replication techniques for variance approximation. In *Proceedings of SAS® Global Forum*, Number 2601–2015. Cary, NC: SAS Institute, Inc. https://support.sas.com/resources/papers/proceedings15/2601-2015.pdf (accessed November 3, 2020).

Lewis, T. H. (2016). *Complex Survey Data Analysis with SAS®.* Boca Raton, FL: CRC Press.

Lohr, S. L. (2012). Using SAS® for the design, analysis, and visualization of complex surveys. In *Proceedings of SAS Global Forum*, Number 343–2012. Cary, NC: SAS Institute, Inc. https://support.sas.com/resources/papers/proceedings12/343-2012.pdf (accessed August 22, 2020).

Lohr, S. L. (2014). Design effects for a regression slope in a cluster sample. *Journal of Survey Statistics and Methodology 2*(2), 97–125.

Lu, Y. and S. L. Lohr (2021). *SDAResources: Datasets and Functions for "Sampling: Design and Analysis."* R package version 0.1.0. `https://CRAN.R-project.org/package=SDAResources` (accessed May 17, 2021).

Lu, Y. and S. L. Lohr (2022). *R Companion for* Sampling: Design and Analysis, 3rd ed. Boca Raton, FL: CRC Press.

Lundquist, P. and C.-E. Särndal (2013). Aspects of responsive design with applications to the Swedish Living Conditions Survey. *Journal of Official Statistics 29*(4), 557–582.

Lydersen, C. and M. Ryg (1991). Evaluating breeding habitat and populations of ringed seals *Phoca hispida* in Svalbard fjords. *Polar Record 27*, 223–228.

Macdonell, W. R. (1901). On criminal anthropometry and the identification of criminals. *Biometrika 1*, 177–227.

Matange, S. (2014). Up your game with graph template language layouts. In *PharmaSUG 2014 Conference Proceedings*. Chapel Hill, NC: PharmaSUG. `https://www.pharmasug.org/2014-proceedings.html` (accessed November 5, 2020).

Matange, S. and J. Bottitta (2016). *SAS® ODS Graphics Designer by Example: A Visual Guide to Creating Graphs Interactively.* Cary, NC: SAS Institute, Inc.

Matange, S. and D. Heath (2011). *Statistical Graphics Procedures by Example: Effective Graphs Using SAS®*. Cary, NC: SAS Institute, Inc.

McConville, K. and F. Breidt (2013). Survey design asymptotics for the model-assisted penalised spline regression estimator. *Journal of Nonparametric Statistics 25*(3), 745–763.

Mukhopadhyay, P. K. (2016). Survey data imputation with PROC SURVEYIMPUTE. In *Proceedings of the SAS® 2017 Global Forum*, Number SAS3520–2016. Cary, NC: SAS Institute, Inc. `https://support.sas.com/resources/papers/proceedings16/SAS3520-2016.pdf` (accessed November 3, 2020).

Mukhopadhyay, P. K., A. An, R. D. Tobias, and D. L. Watts (2008). Try, try again: Replication-based variance estimation methods for survey data analysis in SAS® 9.2. In *Proceedings of SAS® Global Forum*, Number 367–2008. Cary, NC: SAS Institute, Inc. `https://support.sas.com/resources/papers/proceedings/pdfs/sgf2008/367-2008.pdf` (accessed February 19, 2021).

Mukhopadhyay, P. K. and A. McDowell (2011). Small area estimation for survey data analysis using SAS® software. In *Proceedings of SAS® Global Forum*, Number 336–2011. Cary, NC: SAS Institute, Inc. `https://support.sas.com/resources/papers/proceedings11/336-2011.pdf` (accessed November 19, 2020).

National Center for Health Statistics (1987). *Vital Statistics of the United States, Volume 3: Marriage and Divorce.* Washington, DC: U.S. Government Printing Office.

Nolan, D. and T. Speed (2000). *Stat Labs: Mathematical Statistics Through Applications.* New York: Springer.

Peart, D. (1994). *Impacts of Feral Pig Activity on Vegetation Patterns Associated with Quercus agrifolia on Santa Cruz Island, California.* Ph.D. dissertation. Tempe, AZ: Arizona State University.

Pratesi, M. (Ed.) (2016). *Analysis of Poverty Data by Small Area Estimation.* Hoboken, NJ: Wiley.

Rao, J. N. K. and I. Molina (2015). *Small Area Estimation, 2nd ed.* Hoboken, NJ: Wiley.

Rao, J. N. K. and A. J. Scott (1981). The analysis of categorical data from complex sample surveys: Chi-squared tests for goodness of fit and independence in two-way tables. *Journal of the American Statistical Association 76*, 221–230.

Rao, J. N. K. and A. J. Scott (1984). On chi-squared tests for multiway contingency tables with cell proportions estimated from survey data. *The Annals of Statistics 12*, 46–60.

Rao, J. N. K., A. J. Scott, and C. J. Skinner (1998). Quasi-score tests with survey data. *Statistica Sinica 8*, 1059–1070.

Rao, J. N. K., C. F. J. Wu, and K. Yue (1992). Some recent work on resampling methods for complex surveys. *Survey Methodology 18*, 209–217.

Rizzo, L. (2014). A rake-trim SAS® macro and its uses at Westat. In *Proceedings of the SAS® 2014 Global Forum*, Number 1627–2014. Cary, NC: SAS Institute, Inc. `https://support.sas.com/resources/papers/proceedings14/1627-2014.pdf` (accessed November 3, 2020).

Roberts, R. J., Q. D. Sandifer, M. R. Evans, M. Z. Nolan-Ferrell, and P. M. Davis (1995). Reasons for non-uptake of measles, mumps, and rubella catch up immunisation in a measles epidemic and side effects of the vaccine. *British Medical Journal 310*, 1629–1632.

Ruggles, S., M. Sobek, T. Alexander, C. A. Fitch, R. Goeken, P. K. Hall, M. King, and C. Ronnander (2004). Integrated Public Use Microdata Series: Version 3.0 [machine-readable database]. `www.ipums/org/usa` (accessed September 17, 2008).

SAS Institute Inc. (2015). Poisson regressions for complex surveys. `https://support.sas.com/rnd/app/stat/examples/SurveyPoisson/surveypoisson.htm` (accessed December 29, 2020).

SAS Institute Inc. (2021). *SAS/STAT® User's Guide.* Cary, NC: SAS Institute Inc. `https://documentation.sas.com/` (accessed April 27, 2021).

Silva, A. R. (2017). %SURVEYGENMOD macro: An alternative to deal with complex survey design for the GENMOD procedure. In *Proceedings of the SAS® Global Forum 2017*, Number 268–2017. Cary, NC: SAS Institute Inc. `https://support.sas.com/resources/papers/proceedings17/0268-2017.pdf` (accessed December 29, 2020).

Silva, P. H. D. and A. R. Silva (2014). A SAS® macro for complex sample data analysis using generalized linear models. In *Proceedings of the SAS® Global Forum 2014*, Number 1657–2014. Cary, NC: SAS Institute Inc. `https://support.sas.com/resources/papers/proceedings14/1657-2014.pdf` (accessed December 29, 2020).

Singh, K. and M. Xie (2010). Bootstrap method. In *International Encyclopedia of Education*, pp. 46–51. Amsterdam: Elsevier.

Slaughter, S. J. and L. D. Delwiche (2004). SAS® macro programming for beginners. In *Proceedings of the SAS® Users Group International*, Number 243-29. Cary, NC: SAS Institute, Inc. `https://support.sas.com/resources/papers/proceedings/proceedings/sugi29/243-29.pdf` (accessed December 12, 2020).

Stockford, D. D. and W. F. Page (1984). Double sampling and the misclassification of Vietnam service. In *Proceedings of the Social Statistics Section*, pp. 261–264. Alexandria, VA: American Statistical Association.

Stroup, W. W., G. A. Milliken, E. A. Claassen, and R. D. Wolfinger (2018). *SAS® for Mixed Models: Introduction and Basic Applications.* Cary, NC: SAS Institute, Inc.

Tillé, Y. and M. Wilhelm (2017). Probability sampling designs: Principles for choice of design and balancing. *Statistical Science 32*(2), 176–189.

Tzavidis, N., L.-C. Zhang, A. Luna, T. Schmid, and N. Rojas-Perilla (2018). From start to finish: A framework for the production of small area official statistics. *Journal of the Royal Statistical Society: Series A 181*(4), 927–979.

U.S. Bureau of the Census (1921). *Fourteenth Census of the United States Taken in the Year 1920.* Washington, DC: U.S. Government Printing Office. `https://www.census.gov/library/publications/1921/dec/vol-01-population.html` (accessed August 4, 2020).

U.S. Bureau of the Census (1995). *1992 Census of Agriculture, Volume 1: Geographic Area Series.* Washington, DC: U.S. Bureau of the Census.

U.S. Census Bureau (1994). *County and City Data Book: 1994.* Washington, DC: U.S. Census Bureau.

U.S. Census Bureau (2006). *Vehicle Inventory and Use Survey—Methods.* Washington, DC: U.S. Census Bureau.

U.S. Census Bureau (2012). *United States Summary, 2010.* Washington, DC: U.S. Census Bureau. `https://www.census.gov/prod/cen2010/cph-2-1.pdf` (accessed October 3, 2020).

U.S. Census Bureau (2019). State population totals: 2010-2019. Table 1. Annual estimates of the resident population for the United States, regions, states, and Puerto Rico: April 1, 2010 to July 1, 2019 (NST-EST2019-01). `https://www.census.gov/data/datasets/time-series/demo/popest/2010s-state-total.html` (accessed August 3, 2020).

U.S. Census Bureau (2020). *Understanding and Using the American Community Survey Public Use Microdata Sample Files: What Data Users Need to Know.* Washington, DC: U.S. Census Bureau.

U.S. Department of Education (2020). College scorecard data. `https://collegescorecard.ed.gov/data/` (accessed August 25, 2020).

U.S. Department of Justice (1989). *Survey of Youth in Custody, 1987, United States computer file, Conducted by Department of Commerce, Bureau of the Census, 2nd ICPSR ed.* Ann Arbor, MI: Inter-University Consortium for Political and Social Research.

Valliant, R. and K. F. Rust (2010). Degrees of freedom approximations and rules-of-thumb. *Journal of Official Statistics 26*(4), 585–602.

Vijayan, K. (1968). An exact πps sampling scheme: Generalization of a method of Hanurav. *Journal of the Royal Statistical Society, Series B 30*, 556–566.

Virginia Polytechnic and State University/Responsive Management (2006). *An Assessment of Public and Hunter Opinions and the Costs and Benefits to North Carolina of Sunday Hunting.* Blacksburg, VA: Virginia Polytechnic and State University.

Wang, J. (2021). The pseudo maximum likelihood estimator for quantiles of survey variables. *Journal of Survey Statistics and Methodology 9*(1), 185–201.

Westfall, P. H., R. D. Tobias, and R. D. Wolfinger (2011). *Multiple Comparisons and Multiple Tests using SAS®*. Cary, NC: SAS Institute, Inc.

Wilk, S. J., W. W. Morse, D. E. Ralph, and T. R. Azarovitz (1977). *Fishes and Associated Environmental Data Collected in New York Bight, June 1974–June 1975.* NOAA Tech. Rep. No. NMFS SSRF-716. Washington, DC: U.S. Government Printing Office.

Woodruff, R. S. (1952). Confidence intervals for medians and other position measures. *Journal of the American Statistical Association 47*, 636–646.

Yang, S. and J. K. Kim (2016). Fractional imputation in survey sampling: A comparative review. *Statistical Science 31*(3), 415–432.

Zhang, C., C. Antoun, H. Y. Yan, and F. G. Conrad (2020). Professional respondents in opt-in online panels: What do we really know? *Social Science Computer Review 38*(6), 703–719.

Zhang, G., F. Christensen, and W. Zheng (2015). Nonparametric regression estimators in complex surveys. *Journal of Statistical Computation and Simulation 85*(5), 1026–1034.

Zhu, M. (2014). Analyzing multilevel models with the GLIMMIX procedure. In *Proceedings of the SAS® Global Forum 2014*, Number 026–2014. Cary, NC: SAS Institute Inc. `https://support.sas.com/resources/papers/proceedings14/SAS026-2014.pdf` (accessed January 21, 2021).

Index